IF THE UNIVERSE IS TEEMING WITH ALIENS ...

WHERE IS EVERYBODY?

IF THE UNIVERSE IS TEEMING WITH ALIENS ...

WHERE IS EVERYBODY?

FIFTY SOLUTIONS TO THE FERMI PARADOX
AND THE PROBLEM OF EXTRATERRESTRIAL LIFE

Stephen Webb

Copernicus Books   IN ASSOCIATION WITH   Praxis Publishing Ltd.

An Imprint of Springer-Verlag

Published in the United States by Copernicus Books,
an imprint of Springer-Verlag New York, Inc.
A member of BertelsmannSpringer Science+Business Media GmbH

Copernicus Books
37 East 7th Street
New York, NY 10003
www.copernicusbooks.com

Library of Congress Cataloging-in-Publication Data
Webb, Stephen.
   If the universe is teeming with aliens . . . where is everybody? : fifty solutions to the Fermi
   paradox and the problem of extraterrestrial life / Stephen Webb.
     p. cm.
   Includes bibliographical references and index.
   ISBN 0-387-95501-1 (acid-free paper)
    1. Fermi paradox. 2. Life on other planets. 3. Fermi, Enrico, 1901–1954. I. Title.
   QB54.W384 2002
    576.8'39—dc21   2002073910

Manufactured in the United States of America.
Printed on acid-free paper.

9 8 7 6 5 4 3 2 1

ISBN 0-387-95501-1   SPIN 10879867

To Heike

CONTENTS

PREFACE

This book is about the Fermi paradox — the contradiction between the apparent absence of aliens, and the common expectation that we should see evidence of their existence. I was fascinated by the paradox when I first met it some 17 years ago, and it fascinates me still. Over those years, many authors (too many to mention here, though their names appear in the reference list at the back of this book) have enthralled me with their writing about the paradox. Their influence upon this work will be clear. I have also discussed the paradox with many friends and colleagues; although they are too numerous to mention individually, I am indebted to them all.

Several people have contributed directly to the writing of this book, and I would like to take this chance to thank them. Clive Horwood of Praxis Publishing, John Watson of Springer-Verlag and Paul Farrell of Copernicus Books have been very supportive of the project; the book would not have been completed had it not been for their advice and encouragement. (I would also like to thank John for sharing his favored resolution of the paradox over an enjoyable working lunch.) Stuart Clark provided many useful comments on an early draft of the manuscript; Bob Marriott and Timothy Yohn caught several errors and solecisms in a later draft (Bob also sent me a list of 101 resolutions of the paradox — 75 of which I agree with); and I am extremely grateful to Steve Gillett for putting me right on many scientific points. (I am, of course, responsible for those errors that remain.) Mareike Paessler was an exceptionally observant and helpful Production Editor. Her painstaking work with Assistant Editor Anna Painter improved the text enormously. Several authors and organizations kindly gave permission to reproduce figures; I am particularly grateful to thank Lora Gordon, Geoffrey Landis, Ian Wall, Susan Lendroth, Reinhard Rachel, Heather Lindsay and Merrideth Miller for help in obtaining suitable figures. I would like to thank David Glasper for sharing his recollections of a childhood incident that affected us both. Finally, of course, I would like to thank my family — Heike, Ron, Ronnie, Peter, Jackie, Emily and Abigail — for their patience. I spent time writing that I should instead have shared with them.

Stephen Webb
Milton Keynes, July 2002

FIGURE CREDITS

I would like to acknowledge the following sources and copyright holders for granting permission to use their images.

Figures 1, 6, 7, 8, 10 and 28 are courtesy of the American Institute of Physics Emilio Segré Visual Archives. Figure 5 is © *The New Yorker* Collection 1950, drawn by Alan Dunn, from cartoonbank.com. All Rights Reserved. Both Figures 9 and 29 are SETI League photographs, reproduced by permission. Figures 13, 15–19, 21, 22, 44–46, 48–52, 57–60 and 70 are courtesy of NASA. Figure 14 is courtesy of the Astronomical Society of the Pacific. Figure 20 is courtesy of Radford University and Lora Gordon. Figure 23 is courtesy of Michael Carroll and the Planetary Organisation. Figure 25 is courtesy of IoP and Miguel Alcubierre Moya. Figure 27 is courtesy of Geoffrey Landis. Figure 30 is courtesy of LIGO. Figure 31 is courtesy of Antares — F. Montanet CCPM/IN2P3/CNRS — Université Mediterranée. Figures 33 and 34 are both courtesy of the SETI Institute, reproduced by permission. Figure 40 is courtesy of Michael Daly, Uniformed Services, University of the Health Sciences. Figure 41 is courtesy of CERN. Figure 43 is © Frederik Ramm

(see http://www.remote.org/frederik/culture/berlin/).

Figure 53 is courtesy of NASA/P. J. T. Leonard, and Figure 54 is courtesy of NASA/Don Davis. Figure 55 is courtesy of NOAA; photographer Michael Van Woert. Figure 56 is courtesy of NOAA. The four images in Figure 62 are courtesy of Prof. Dr. K. O. Stetter and Dr. R. Rachel, Universität Regensburg, Mikrobiologie; © University of Regensburg. Figure 64 is courtesy of the Wellcome Trust. Figure 67 is © US Department of Energy Human Genome Project. Figure 69 is © Arizona State University, photograph by Alan Riggs. Figure 71 is courtesy of Creswell Crags, © Creswell Heritage Trust.

1

Where *Is* Everybody?

There is something beguiling about paradox. The impossible and paradoxical prints of Maurits Escher never fail to deceive the eye. Poems like Robert Graves' *Warning to Children*, which play with the paradox of infinite regress, make the head spin. Paradox lies at the heart of Joseph Heller's *Catch-22*, one of the 20th century's greatest novels. My favorite paradox, though, is that of Fermi.

I first came across the Fermi paradox in the summer of 1984. I had just graduated from Bristol University, and I should have spent the summer months studying Aitchison and Hey's *Gauge Theories in Particle Physics* — required reading before I started postgraduate studies at Manchester University. Instead, I spent my time enjoying the sunshine on the Bristol Downs, studying my favorite reading matter: *Isaac Asimov's Science Fiction Magazine*. (As with many people, SF sparked my interest in science. It was through reading the works of Isaac Asimov, Arthur Clarke and Robert Heinlein and watching films like *Forbidden Planet* that I became enamored with science.[1]) Two thought-provoking science-fact articles appeared in successive issues of *IASFM* that year. The first, by Stephen Gillett, was simply entitled *The Fermi Paradox*. The second, a forceful rebuttal by Robert Freitas, was entitled *Fermi's Paradox: A Real Howler*.[2]

Chapter 1

Gillett argued in the following way. Suppose, as the optimists believed, that the Galaxy is home to many extraterrestrial civilizations. (To save typing, I shall often refer to an extraterrestrial civilization as an ETC.) Then, since the Galaxy is extremely old, the chances are good that ETCs will be millions or even *billions* of years in advance of us. The Russian astrophysicist Nikolai Kardashev proposed a useful way of thinking about such civilizations. He argued that ETCs would possess one of three levels of technology. A Kardashev type 1 civilization, or K1 civilization, would be comparable to our own: it could employ the energy resources of a planet. A K2 civilization would be beyond our own: it could employ the energy resources of a star. A K3 civilization could employ the energy resources of an entire *galaxy*. According to Gillett, then, most ETCs in the Galaxy would be of a K2 or K3 type. Now, everything we know about terrestrial life tells us that life has a natural tendency to expand into all available space. Why should extraterrestrial life be any different? Surely ETCs would want to expand from their home world and out into the Galaxy. The key point, however, is that a technologically advanced ETC could colonize the Galaxy in a few million years. They should already be here! The Galaxy *should* be swarming with life. And yet we see no evidence that ETCs exist. Gillett called this the Fermi paradox. (I learned why Fermi's name is attached to the paradox a few months later, when Eric Jones published a Los Alamos preprint describing the origins of the paradox; but more of this later.) For Gillett, the paradox pointed to a chilling conclusion: mankind is alone in the Universe.

Freitas thought this was all hogwash. He compared Gillett's logic to the following argument: Lemmings breed quickly — about 3 litters per year, with each litter containing up to 8 offspring. In just a few years the total mass of lemmings will be equal to the mass of the entire terrestrial biosphere. The Earth must be swarming with lemmings. And yet, most of us see no evidence that lemmings exist. Have *you* ever seen a lemming? The "Fermi paradox" line of reasoning would lead us to conclude that lemmings do not exist — yet, as Freitas pointed out, this would be absurd. More interestingly, he believed the lack of evidence for ETCs is not particularly strong: if small artificial probes were parked in the Asteroid Belt, say, or larger probes in the Oort Cloud, then we would have no chance of detecting them. Besides, he argued that the logic behind the so-called paradox is faulty. The first two steps in the argument are: (i) if aliens exist, then they should be here; (ii) if they are here, then we should observe them. The difficulty is those two "should"s. A "should" is not a "must," and therefore it is logically incorrect to reverse the arrow of implication. (In other words, the fact we have not observed them does not allow us to conclude they are not here, so we cannot conclude they do not exist.)

Until there is clear evidence to resolve a paradox, people are free to follow different lines of reasoning. This is what makes a paradox so interesting. In the case of the Fermi paradox, the stakes are so high (the existence or otherwise of alien intelligence) and the experimental input to the argument is so sparse (even now, we cannot be sure ETCs are not here) that arguments often become heated. In the Gillett–Freitas debate, I initially sided with Freitas. The main reason was sheer weight of numbers: there are perhaps as many as 400 billion stars in the Galaxy, and as many galaxies in the Universe as there are stars in the Galaxy. Ever since Copernicus, science has taught us there is nothing special about Earth. It followed, then, that Earth could not be the sole home to intelligent life. And yet ...

I could not get Gillett's argument out of my mind. I had been reading about cosmic wonders since I was a child. The Galaxy-spanning civilization of the *Foundation* trilogy, the astroengineering wonders of *Ringworld*, the enigma of the vessel in *Rendezvous with Rama* — all these were part of my mental furniture. And yet where *were* these marvels? The imaginations of SF writers had shown me hundreds of possible universes, but my astronomy lecturers made it clear that so far, whenever we look out into the real Universe, we can explain everything we see in terms of the cold equations of physics. Put simply, the Universe looks dead. The Fermi question: where *is* everybody? The more I thought about it, the more the paradox seemed to be significant.

*   *   *

It seemed to me the paradox was a competition between two large numbers: the vast number of potential sites for life versus the vast age of the Universe.

The first number is simply the number of planets with suitable environments for the development of life. If we adopt the Principle of Mediocrity, and assume there is nothing at all about special about Earth, then it follows there are many millions of suitable environments for life in the Galaxy (and many billions of environments in the Universe). Given so many potential seeding grounds, life should be common.

The second number is simply the age of the Universe: the latest measurements suggest it is slightly more than 13 billion years old. To evoke a feeling for such a large time span, it is usual in these discussions to compress the entire history of the Universe into a standard length or interval. In this case, I will compress the current age of the Universe into a standard Earth year: in other words, the "Universal Year" compresses the entire history of the Universe into 365 days. On this timescale, a second of real time corresponds to 400 years; in other words, in the Universal Year, western science begins about 1 second before midnight on 31 December. The whole

3

history of our species is much less than 1 hour of the Universal Year. The earliest ETCs, however, could have originated in the early summer months of the Universal Year. If the colonization of the Galaxy can take place in the equivalent of a few hours, then one would expect one or more of the advanced technological civilizations to have long since completed the job. At the very least, if they really were so far beyond us, one would expect to see or hear *some* evidence of their presence. But the Universe is silent. The Fermi paradox might not logically *prove* aliens do not exist, but surely it is a problem demanding solution.

TABLE 1   In the "Universal Year," we compress 13 billion years into 365 days.

"Real" time	Time in a Universal Year
50 yrs	0.125 s
100 yrs	0.25 s
400 yrs	1 s
1000 yrs	2.5 s
2000 yrs	5 s
10 000 yrs	25 s
100 000 yrs	4 mins 10 s
1 million yrs	41 mins 40 s
2 million yrs	1 hr 23 mins 20 s
10 million yrs	6 hr 56 min 40 s
100 million yrs	2 days 21 hr 26 min 40 s

I was not the only one who found the Fermi paradox interesting. Over the years, many people have offered their resolutions to the paradox, and I developed the habit of collecting them. Although there is a fascinating range of answers to the question "where is everybody?," they all fall into one of three classes.

First, there are answers based around the idea that somehow the extraterrestrials are (or have been) here. This is probably the most popular resolution of the paradox. Certainly, belief in intelligent extraterrestrial life is widespread. In a CNN Internet poll on 1 July 2000, of the 6399 people who voted, 82% thought there is intelligent life elsewhere in the Universe. As of the 2001 summer solstice, 94% of the 94,319 respondents to a SETI@home poll believe life exists outside Earth. Several polls suggest the majority of American people believe flying saucers exist and are here; the proportion of believers seems to be less among Europeans, but is nevertheless high.

Second, there are answers suggesting ETCs exist, but for some reason we have not yet found evidence of their existence. This is probably the most popular category of answer among practicing scientists.

Third, there are answers purporting to explain why mankind is alone in the Universe, or at least in the Galaxy; we do not hear from extraterrestrial intelligence because there *is* no extraterrestrial intelligence.

The purpose of this book is to present and discuss 50 proposed solutions to Fermi's question. The list is not intended to be exhaustive; rather, I have chosen them because they are representative (and also because I think they are particularly interesting). The proposed solutions come from scientists working in several widely separated fields of science, but also from SF

4

authors; in this topic, authors have been at least as industrious as scientists, and in many cases they have anticipated the work of scientists.

The outline of the book is as follows.

Chapter 2 gives a brief biography of Fermi, focusing on his scientific achievements; I then discuss the notion of paradox and present a brief discussion of the history of the Fermi paradox.

Chapters 3–5 present 49 of my favorite solutions to the paradox; not all of them are independent, and sometimes I revisit a solution in another guise, but all of them have been seriously proposed as answering Fermi's question. I arrange the answers according to the three classes mentioned above: Chapter 3 contains answers based around the idea that ETCs are here; Chapter 4 contains answers based around the idea that ETCs exist, but we have not yet found evidence of them; Chapter 5 contains answers based around the idea that we are alone. There is a logic to the arrangement of the solutions, but I hope the discussions are self-contained enough to allow readers to "dip into" the book and pick out solutions that particularly interest them. In the discussions I will try to be as even-handed as possible, even if I disagree with the solution (which I often do).

Chapter 6 contains the 50th solution: my own view of the resolution of the paradox. It is not an original suggestion, but it summarizes what I feel the Fermi paradox can tell us about the Universe in which we live.

Superscripted numbers, which appear throughout the book, refer to numbered items in Chapter 7; these items contain notes and suggestions for further reading. Since the material in this book covers a wide range of subjects from astronomy to zoology, and since the space for the discussions is necessarily limited (it works out at an average of about 5 pages per solution), I have also given a wide-ranging list of references. The references themselves, which are referred to in Chapter 7 by numbers in square brackets, appear in Chapter 8. They range from SF stories to primary research articles in scholarly journals. Many readers may find it difficult to access the more specialized references, but I hope they will at least find it possible to use these references to help find related information on the Web.

The book is specifically aimed at a popular audience. One of the beauties of the Fermi paradox is that it can be appreciated without the need for any mathematics beyond an understanding of exponential notation.[3] It follows that anyone can present a resolution of the Fermi paradox; you do not need to have years of scientific and mathematical training to contribute to the debate. (Indeed, as I noted above, many of the best ideas have come from SF writers rather than scientists.) I hope that a reader of this book may devise a solution that no-one else has thought of. If you do — please write to me and share it!

2

Of Fermi
and Paradox

Before looking at the various proposed solutions to the Fermi paradox, this chapter presents some of the background. I first give a short
biography of Enrico Fermi himself, focusing on just a few of his scientific accomplishments (those that I will refer to in later sections of the
book). Fermi led an interesting life outside of science, though, and I recom-
mend the interested reader to the biographies of Fermi listed in Chapter 7.
I then discuss the notion of paradox, and briefly look at a few examples
from various fields. Paradox has played an important role in intellectual
history, helping thinkers to widen their conceptual framework and sometimes forcing them to accept quite counterintuitive notions. It is interesting
to compare the Fermi paradox with these more established paradoxes. Finally, I discuss how Fermi's name came to be attached to a paradox that is
older than many people believe.

Chapter 2

ENRICO FERMI

It is no good to try to stop knowledge from going forward.
Ignorance is never better than knowledge.
Enrico Fermi

Enrico Fermi was the most complete physicist of the last century — a world-class theoretician who carried out experimental work of the highest order. No other physicist since Fermi has switched between theory and experiment with such ease, and it is unlikely that anyone will do so again. The field has become too large to permit such crossover.

Fermi was born in Rome on 29 September 1901, the third child of Alberto Fermi, a civil servant, and Ida DeGattis, a schoolteacher. He showed precocious ability in mathematics,[4] and as an undergraduate student of physics at the Scuola Normale Superiore in Pisa he quickly outstripped his teachers.[5]

His first major contribution to physics was an analysis of the behavior of certain fundamental particles that make up matter. (These particles — such as protons, neutrons and electrons — are now called *fermions* in his honor.) Fermi showed that, when matter is compressed so that identical fermions are brought close together, a repulsive force comes into play that resists further compression. This fermionic repulsion plays an important role in our understanding of phenomena as diverse as the thermal conductivity of metals and the stability of white dwarf stars.

Soon after, Fermi's theory of beta decay (a type of radioactivity in which a massive nucleus emits an electron) cemented his international reputation. His theory demanded that a ghostly particle be emitted along with the electron, a particle he called the *neutrino* — "little neutral one." Not everyone believed in the existence of this hypothetical fermion, but Fermi was proved correct. Physicists finally detected the neutrino in 1956. Although the neutrino remains rather ghostly in its reluctance to react with normal matter, its properties play a profound role in present-day astronomical and cosmological theories.

In 1938, Fermi won the Nobel Prize for physics. The award was partly in recognition of a technique he developed to probe the atomic nucleus. His technique led him to the discovery of new radioactive elements; by bombarding the naturally occurring elements with neutrons, he produced more than 40 artificial radioisotopes. The award also recognized his discovery of how to make neutrons move slowly. This may seem like a minor discovery, but it has profound practical applications, since slow-moving neutrons are more effective than fast neutrons at inducing radioactivity. (A slow neutron spends more time in the neighborhood of a target nucleus, and so is more likely to interact with the nucleus. In a similar way, a well-aimed golf ball

FIGURE 1 *This photograph of Enrico Fermi lecturing on atomic theory appears on a 34¢ stamp, released by the US Postal Service on 29 September 2001 to commemorate the hundredth anniversary of Fermi's birth.*

is more likely to sink into the hole if it is moving slowly: a fast-moving putt can roll by.) This principle is used in the operation of nuclear reactors.

News of the award was tempered by the worsening political situation in Italy. Mussolini, increasingly influenced by Hitler, initiated an anti-Semitic campaign. Italy's fascist government passed laws that were copied directly from the Nazi Nuremberg edicts. The laws did not directly affect Fermi or his two children, who were considered to be Aryans, but Fermi's wife, Laura, was Jewish. They decided to leave Italy, and Fermi accepted a position in America.

Two weeks after arriving in New York, news reached Fermi that German and Austrian scientists had demonstrated nuclear fission. Einstein, after some prompting, wrote his historic letter to Roosevelt alerting the President to the probable consequences of nuclear fission. Citing work by

Fermi and his colleagues, Einstein warned that a nuclear chain reaction might be set up in a large mass of uranium — a reaction that could lead to the release of vast amounts of energy. Roosevelt was concerned enough to fund a program of research into the defense possibilities. Fermi was deeply involved in the program.

Physicists had many questions to answer before they could build a bomb, and it was Fermi who answered many of them. On 2 December 1942, in a makeshift laboratory constructed in a squash court under the West Stands of the University of Chicago stadium, Fermi's group successfully achieved the first self-sustaining nuclear reaction. The reactor, or pile, consisted of slugs of purified uranium — about 6 tons in all — arranged within a matrix of graphite. The graphite slowed the neutrons, enabling them to cause further fission and maintain the chain reaction. Control rods made of cadmium (a strong neutron absorber) controlled the rate of the chain reaction. The pile went critical at 2:20 P.M., and the first test was run for 28 minutes.[6]

Fermi, with his unmatched knowledge of nuclear physics, played an important role in the Manhattan Project. He was there in the Alamogordo desert on 15 July 1945, 9 miles away from ground zero at the Trinity test. He lay on the ground facing in the direction opposite the bomb. When he saw the flash from the immense explosion, he got to his feet and dropped small pieces of paper from his hand. In still air the pieces of paper would have fallen to his feet; but when the shock wave arrived, a few seconds after the flash, the paper moved horizontally due to the displacement of air. In typical fashion, he measured the displacement of the paper; since he knew the distance to the source, he could immediately estimate the energy of the explosion.

After the war, Fermi returned to academic life at the University of Chicago and became interested in the nature and origin of cosmic rays. In 1954, however, he was diagnosed with stomach cancer. Emilio Segré, Fermi's lifelong friend and colleague, visited him in hospital. Fermi was resting after an exploratory operation, and was being fed intravenously. Even at the end, according to Segré's touching account, Fermi retained his love of observation and calculation: he measured the flux of the nutrient by counting drops and timing them with a stopwatch.

Fermi died on 29 November 1954, at the early age of 53.

Fermi Questions

Fermi's colleagues prized him for his uncanny ability to see straight to the heart of a physical problem and describe it in simple terms. They called him the Pope, because he seemed infallible. Almost as impressive was the way he estimated the magnitude of an answer (often by doing complex calculations in his head). Fermi tried to inculcate this facility in his students. He would demand of them, without warning, answers to seemingly unanswerable questions. How many grains of sand are there on the world's beaches? How far can a crow fly without stopping? How many atoms of Caesar's last breath do you inhale with each lungful of air? Such "Fermi questions" (as they are now known) required students to draw upon their understanding of the world and their everyday experience and make rough approximations, rather than rely on bookwork or prior knowledge.

The archetypal Fermi question is one he asked his American students: "How many piano tuners are there in Chicago?" We can derive an informed estimate, as opposed to an uninformed guess, by reasoning as follows. First, suppose that Chicago has a population of 3 million people. (I have not checked an almanac to see whether this is correct; but making explicit estimates in the absence of certain knowledge is the whole point of the exercise. Chicago is a big city, but not the biggest in America, so we can be confident that the estimate is unlikely to be in error by more than a factor of 2. Since we have explicitly stated our assumption we can revisit the calculation at a later date, and revise the answer in the light of improved data.) Second, assume that families, rather than individuals, own pianos and ignore those pianos belonging to institutions like schools, universities and orchestras. Third, if we assume that a typical family contains 5 members, then our estimate is that there are 600,000 families in Chicago. We know that not every family owns a piano; our fourth assumption is that 1 family in 20 owns a piano. We thus estimate there are 30,000 pianos in Chicago. Now ask the question: How many tunings would 30,000 pianos require in 1 year? Our fifth assumption is that a typical piano will require tuning once per year — so 30,000 piano tunings take place in Chicago each year. Assumption six: A piano tuner can tune 2 pianos per day and works on 200 days in a year. An individual piano tuner therefore tunes 400 instruments in 1 year. In order to accommodate the total number of tunings required, Chicago must be home to $30,000/400 = 75$ piano tuners. We want an estimate, not a precise figure, so finally we round this number up to an even 100.

As we shall see later, Fermi's ability to grasp the essentials of a problem manifested itself when he posed the question: "Where *is* everybody?"

Chapter 2

PARADOX

These are old fond paradoxes, to make fools laugh i' the alehouse.
William Shakespeare, *Othello*, Act II, Scene I

Our word *paradox* comes from two Greek words: *para*, meaning "contrary to," and *doxa*, meaning "opinion."[7] It describes a situation in which, alongside one opinion or interpretation, there is another, mutually exclusive opinion. The word has taken on a variety of subtly different meanings, but at the core of each usage is the idea of a contradiction. Paradox is more than mere inconsistency, though. If you say "it is raining, it is not raining," then you have contradicted yourself, but paradox is more than this. A paradox arises when you begin with a set of self-evident premises and then, from these premises, deduce a conclusion that undermines them. If you have a cast-iron argument that proves it *must* certainly be raining outside, and then you look out of the window and see that it is *not* raining, then you have a paradox to resolve.

A weak paradox or *fallacy* can often be clarified with a little thought. The contradiction usually arises because of a simple mistake in a chain of logic leading from premises to conclusion.[8] In a strong paradox, however, the source of a contradiction is not immediately apparent; centuries may pass before matters are resolved. A strong paradox has the power to challenge our most cherished theories and beliefs. Indeed, as the mathematician Anatol Rapoport once remarked: "Paradoxes have played a dramatic part in intellectual history, often foreshadowing revolutionary developments in science, mathematics and logic. Whenever, in any discipline, we discover a problem that cannot be solved within the conceptual framework that supposedly should apply, we experience shock. The shock may compel us to discard the old framework and adopt a new one."[9]

FIGURE 2   *A visual paradox. These impossible figures are Penrose triangles. They appear to show a three-dimensional triangular solid, but these triangles are impossible to construct. Each vertex of a Penrose triangle is in fact a perspective view of a right angle. Artists like Escher delight in presenting visual paradoxes.*

Paradoxes abound in logic and mathematics and physics, and there is a type for every taste and interest.

12

A Few Logical Paradoxes

An old paradox, contemplated by philosophers since the middle of the 4th century BC and still much discussed, is the liar paradox. Its most ancient attribution is to Eubulides of Miletus, who asked: "A man says that he is lying; is what he says true or false?" However one analyzes the sentence, there is a contradiction. The same paradox appears in the New Testament. St. Paul, referring to Cretans, wrote: "One of themselves, even a prophet of their own, said the Cretans are always liars."[10] It is not clear whether St. Paul was aware of the problem in his sentence, but when self-reference is allowed paradox seems almost inevitable.

One of the most important tools of reasoning we possess is the *sorites*. In logicians' parlance, a sorites is a chain of linked syllogisms: the predicate of one statement becomes the subject of the following statement. The following is a typical example:

all ravens are birds;
all birds are animals;
all animals require water to survive.

Following the chain, we reach a logical conclusion: all ravens need water.

Sorites are important because they allow us to make conclusions without covering every eventuality in an experiment. (So we do not need to deprive ravens of water to know they may die of thirst.) But sometimes the conclusion of a sorites can be absurd: we have a sorites paradox. For example, if we accept that adding one grain of sand to another grain of sand does not make a heap of sand, and given that a single grain does not itself constitute a heap, then we must conclude that no amount of sand can make a heap. And yet we see heaps of sand. The source of such paradoxes lies in the intentional vagueness of a word like "heap"; politicians, of course, routinely take advantage of these linguistic tricks.[11]

As well as sorites, when reasoning we all routinely employ induction — the drawing of generalizations from specific cases. For example, whenever we see something drop, it falls *down*: using induction we propose a general law, namely that when things drop they *always* fall down and never up. Induction is such a useful technique that anything casting doubt on it is troubling. Consider Hempel's raven paradox.[12] Suppose that an ornithologist, after years of field observation, has observed hundreds of black ravens. The evidence is enough for her to suggest the hypothesis that "all ravens are black." This is the standard process of scientific induction. Every time the ornithologist sees a black raven it is a small piece of evidence in favor of her hypothesis. Now, the statement that "all ravens are black" is logically equivalent to the statement that "all non-black things are non-ravens." If

13

the ornithologist sees a piece of white chalk, then the observation is a small piece of evidence in favor of the hypothesis that "all non-black things are non-ravens" — but therefore it must be evidence for her claim that ravens are black. Why should an observation regarding chalk be evidence for a hypothesis regarding birds? Does it mean that ornithologists can do valuable work whilst sat indoors watching television, without bothering to watch a bird in the bush?

Another paradox in logic is that of the unexpected hanging, wherein a judge tells a condemned man: "You will hang one day next week but, to spare you mental agony, the day that the sentence will be carried out will come as a surprise." The prisoner reasons that the hangman cannot wait until Friday to carry out the judge's order: so long a delay means everyone will know the execution takes place that day — the execution will not come as a surprise. So Friday is out. But if Friday is ruled out, Thursday is ruled out by the same logic. Ditto Wednesday, Tuesday and Monday. The prisoner, mightily relieved, reasons that the sentence cannot possibly take place. Nevertheless, he is completely surprised as he is led to the gallows on Thursday! This argument — which also goes under the name of the "surprise examination paradox" and the "prediction paradox" — has generated a huge literature.[13]

A Few Scientific Paradoxes

Although it is often fun, and occasionally useful, to ponder liars, ravens and hanged men, arguments involving logical paradoxes too frequently — for my taste at least — degenerate into a discussion over the precise meaning and usage of words. Such discussions may be fine if one is a philosopher. But for my money the really fascinating paradoxes are those that can be found in science.

Consider one of the oldest of all paradoxes: Zeno's paradox of Achilles and the tortoise.[14] Achilles and the tortoise take part in a 100-m sprint. Since Achilles runs 10 times faster than the tortoise, he gives the animal a head start of 10 m. The two sprinters set off at the same instant; so when Achilles has covered the first 10 m, the tortoise has moved on by 1 m. In the time it takes Achilles to cover 1 m, the tortoise has moved on by 10 cm; in the time it takes Achilles to cover that 10 cm, the tortoise has moved on by a further 1 cm. And so on *ad infinitum*. Our senses tell us a fast runner will always overtake a slow runner, but Zeno said Achilles cannot catch the tortoise. There is a contradiction between logic and experience: there is a paradox. It took 2000 years to resolve the paradox — but the mathematical machinery for doing so found a host of other uses.[15]

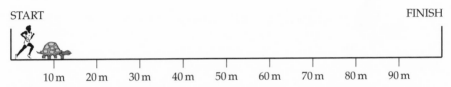

FIGURE 3   When the race begins, Achilles is 10 m behind the tortoise. By the time Achilles has run 10 m, the tortoise has crawled a distance of 1 m. By the time Achilles has run a further 1 m, the tortoise has crawled a further 10 cm. Following this logic, it seems Achilles can never catch up. . . .

The twin paradox, which involves the special relativistic phenomenon of time dilation, is one of the most famous in physics. Suppose one twin stays at home while the other twin travels to a distant star at close to the speed of light. To the stay-at-home twin, his sibling's clock runs slow: his twin ages more slowly than he does. Although this phenomenon may be contrary to common sense, it is an experimentally verified fact. But surely relativity tells us that the traveling twin can consider himself to be at rest? From *his* point of view, the clock of the earthbound twin runs slow; the stay-at-home twin should be the one who ages slowly. So what happens when the traveler returns? They cannot both be right: it is impossible for *both* twins to be younger than each other! The resolution of this paradox is easy: the confusion arises from a simple misapplication of relativity. The twins' situations are not interchangeable: the traveling twin accelerates to light speed, decelerates at the half-way point of his journey, and does it all again on the trip back. Both twins agree that the stay-at-home twin undergoes no such acceleration. So the traveler ages more slowly than the earthbound twin; he returns to find his brother aged, or even dead. An extraterrestrial visitor to Earth would observe the same phenomenon when it returned to its home planet: its stay-at-home siblings (if aliens have siblings) would be older or long-since dead. It is a sad fact of interstellar travel, and it is contrary to our experience, but it is not a paradox.[16]

One of the most important of scientific paradoxes is that named after Heinrich Olbers.[17] He considered a question asked by countless children — "Why is the night sky dark?" — and showed that the darkness of night is deeply mysterious. His reasoning was based upon two premises. First, that the Universe is infinite in extent. Second, that the stars are scattered randomly throughout the Universe. (Olbers did not know of the existence of galaxies — they were not recognized as stellar groupings until some 75 years after his death — but this does not affect his reasoning. His argument works in exactly the same way for galaxies as it does for stars.) From these premises we reach an uncomfortable conclusion: in whichever direction you look, your line-of-sight must eventually end on a star — the night sky should therefore be bright.

Olbers' Paradox

Suppose all stars have the same intrinsic brightness. (The following argument is simpler under this assumption, but the conclusion in no way depends upon it.) Now consider a thin shell of stars (call it shell A) with Earth at its center, and another thin shell of stars (shell B), also centered on Earth, with a radius twice that of shell A. In other words, shell B is twice as distant from us as shell A.

A star in shell B will appear to be $\frac{1}{4}$ as bright as a star in shell A. (This is the inverse-square law: if the distance to a light source *increases* by a factor of 2, the apparent brightness of the light source *decreases* by a factor of $2 \times 2 = 4$.) On the other hand, the surface area of shell B is 4 times larger than that of shell A, so it contains 4 times as many stars. Four times as many stars, each of which is $\frac{1}{4}$ as bright: the total brightness of shell B is exactly the same as the total brightness of shell A! But this works for *any* two shells of stars. The contribution to the brightness of the night sky from a distant shell of stars is the same as from a nearby shell. If the Universe is infinite in extent, then the night sky should be infinitely bright.

This argument is not quite correct: the light from an extremely distant star will be intercepted by an intervening star. Nevertheless, in an infinite Universe with a uniform distribution of stars *any* line of sight will eventually run into a star. Far from being dark, the entire night sky should be as dazzling as the Sun. The night sky should blind us with its brightness!

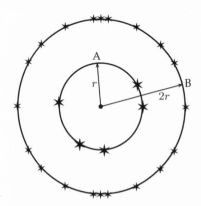

FIGURE 4   *If the stars are uniformly distributed throughout space, then shell B will contain 4 times as many stars as shell A (A is at a distance r and B is at distance 2r). But the stars in shell A will appear 4 times as bright as the stars in shell B. So the total brightness of the shells will be the same. Since there is an infinite number of such shells, the night sky should be infinitely bright. Even allowing for the stars in nearby shells that block the light from distant stars, the night sky should be blindingly bright.*

How can we resolve the paradox? The first explanation you are likely to think of is that clouds of gas or dust obscure the light from distant stars. The Universe does indeed contain dust clouds and gaseous regions, but they cannot shade us from Olbers' paradox: if the clouds absorb light, they will heat up until they are at the same average temperature as the stars themselves. It turns out that the paradox is explained by one of the most dramatic discoveries made by astronomers: the Universe has a finite age. Since the Universe is only about 13 billion years old, the part that we can see is only about 13 billion light years in size. For the night sky to be as bright as the surface of the Sun, the observable Universe would have to be almost 1 million times bigger than it is. (That the Universe is expanding also helps to explain the paradox: light from distant objects is redshifted by the expansion, and so distant objects are less bright than one would expect from the inverse-square law. The principal explanation, though, comes from the finite age of the Universe.)

It is fascinating that in pondering such a simple question — "Why is the night sky dark?" — one could infer that the Universe is expanding and that it (or at least the stars and galaxies it contains) has a finite age. Perhaps the simple question that Fermi asked — "Where *is* everybody?" — leads to an even more important conclusion.

THE FERMI PARADOX

> *Sometimes I think we're alone. Sometimes I think we're not.*
> *In either case, the thought is staggering.*
> Buckminster Fuller

Thanks to detective work by the Los Alamos physicist Eric Jones, whose report I draw heavily upon in this section, we know the genesis of the Fermi paradox.[18]

*   *   *

The spring and summer of 1950 saw the New York newspapers exercised over a minor mystery: the disappearance of public trash cans. This year was also the height of flying saucer reports, another subject that filled the column inches. On 20 May 1950, *The New Yorker* published a cartoon by Alan Dunn that made amusing reference to both stories.

Fermi was at Los Alamos in the summer of 1950. One day, he was chatting to Edward Teller and Herbert York as they walked over to Fuller Lodge for lunch. Their topic was the recent spate of flying saucer observations. Emil Konopinski joined them and told them of the Dunn cartoon. Fermi remarked wryly that Dunn's was a reasonable theory because it accounted for two distinct phenomena: the disappearance of trash cans and the reports of flying saucers. After Fermi's joke, there followed a serious discussion about whether flying saucers could exceed the speed of light. Fermi

FIGURE 5   *For reasons that make sense only to them, aliens are returning to their home planet with trash cans that are the property of New York's Department of Sanitation.*

asked Teller what he thought the probability might be of obtaining evidence for superluminal travel by 1960. Fermi said that Teller's estimate of one-in-a-million was too low; Fermi thought it was more like one-in-ten.

The four of them sat down to lunch, and the discussion turned to more mundane topics. Then, in the middle of the conversation and out of the clear blue, Fermi asked: "Where *is* everybody?" His lunch partners Teller, York and Konopinski immediately understood that he was talking about extraterrestrial visitors. And since this was Fermi, perhaps they realized that it was a more troubling and profound question than it first appears. York recalls that Fermi made a series of rapid calculations and concluded that we should have been visited long ago and many times over.

Although neither Fermi nor the others ever published any of these calculations, we can make a reasonable guess at his thought processes. He must first have made an estimate of the number of ETCs in the Galaxy, and this is something we can estimate ourselves. After all, the question "How many advanced communicating extraterrestrial civilizations are there in the Galaxy?" is a typical Fermi question!

FIGURE 6   *Edward Teller (left) with Fermi in 1951, not long after Fermi first asked his question.*

A Fermi Question: How Many Communicating Civilizations Exist?

Represent the number of communicating ETCs in the Galaxy by the symbol N. To estimate N we first need to know the yearly rate R at which stars form in the Galaxy. We also need to know the fraction f_p of stars that possess planets and, for planet-bearing stars, the number n_e of planets with environments suitable for life. We also need the fraction f_l of suitable planets on which life actually develops; the fraction f_i of these planets on which life develops intelligence; and the fraction f_c of intelligent life-forms that develop a culture capable of interstellar communication. Finally, we need to know the time L, in years, that such a culture will devote to communication. Multiplying all these factors together will provide us with an estimate for N. We can write it as a simple equation:

$$N = R \times f_p \times n_e \times f_l \times f_i \times f_c \times L.$$

The equation $N = R \times f_p \times n_e \times f_l \times f_i \times f_c \times L$ is no more a "proper" equation for the number of communicating ETCs than $N = p_c \times n_f \times f_p \times n_t \times R$ is the equation for the number of piano tuners in Chicago. But if we assign reasonable values to the various factors in the equation — always with the understanding that such values can and will change as our knowledge

FIGURE 7   *Herbert York, one of Fermi's lunchtime companions.*

increases — we will arrive at an estimate for the number of ETCs in the Galaxy. The difficulty we face is in our varying degrees of ignorance for the various terms in the equation. When asked to provide values for these terms, astronomers would provide responses ranging from "We're reasonably certain" (for the factor R) to "We're close to pinning it down" (for the factor f_p) to "How the hell should we know?" (for the factor L). At least when we try to estimate the number of Chicago-based piano tuners, we can be reasonably confident that our various sub-estimates are not wildly in error; there can be no such confidence with our estimate for the number of communicating ETCs. Nevertheless, in the absence of any definite knowledge of ETCs, it is our only way to proceed. (The equation above has reached a certain iconic status in science; it is known as the *Drake equation*, after the radio astronomer Frank Drake who was the first to make explicit use of it.[19]   The Drake equation was the focal point of an extremely influential conference on the search for extraterrestrial intelligence, held at Green Bank in 1961 — 11 years after Fermi's remark.)

   In 1950, Fermi would have known far less about the various factors in the above "equation," but he could have made some reasonable guesses

FIGURE 8   *Emil Konopinski (far left), another one of Fermi's lunchtime companions.*

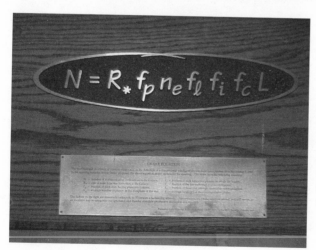

FIGURE 9   *The Drake equation is a means of estimating the number of communicative civilizations in the Galaxy. Drake developed the equation so that it could form the agenda for the first ever SETI meeting (held at NRAO Green Bank, WV, in 1961). This commemorative plaque is on the same wall that held the blackboard where the equation was first written.*

— guided, as he would have been, by the *Principle of Mediocrity*: there is nothing special about Earth or our Solar System. If he guessed at a rate of star formation of 1 star per year he would not have been too wrong. Values of $f_p = 0.5$ (half the stars have planets) and $n_e = 2$ (stars with planets on average each have 2 planets with environments conducive to life) seem to be "reasonable." The other factors are much more subjective; if he were an optimist, Fermi might have chosen $f_l = 1$ (every planet that can develop life *will* develop life), $f_i = 1$ (once life develops, intelligent life will certainly follow), $f_c = 0.1$ (1 in 10 intelligent life-forms will develop a civilization capable and willing to communicate) and $L = 10^6$ (civilizations remain in the communication phase for about 1 million years). Had he argued like that, he would have arrived at the estimate $N = 10^6$. In other words, there could right now be a million civilizations trying to communicate with us. So why do we not hear from some of them? In fact, why are they not already here? If some of the civilizations are extremely long-lived, then we might expect them to colonize the Galaxy — and have done so before multicellular life even developed on Earth. The Galaxy should be swarming with extraterrestrial civilizations. Yet we see no sign of them. We should already know of their existence, but we do not. Where is everybody? *Where are they?* This is the Fermi paradox.

Note that the paradox is not that extraterrestrial intelligence does not exist. (I do not know whether Fermi believed in the existence of extraterrestrial intelligence, but I suspect that he did.) Rather, the paradox is that we see no signs of such intelligence when we might expect to. One explanation of the paradox is indeed that we are the only advanced civilization — but it is only one of several explanations.

*   *   *

Asking why we see no evidence of extraterrestrial civilizations may seem like a trivial question but, as we might expect from a remark by Fermi, it is a profound puzzle. We can appreciate the strength of the paradox when we realize that it has been independently discovered *four* times: it might more properly be called the Tsiolkovsky–Fermi–Viewing–Hart paradox.

Konstantin Tsiolkovsky, a scientific visionary who worked out the theoretical basis of spaceflight as long ago as 1903, believed deeply in the monistic doctrine that ultimate reality is entirely of one substance. If all parts of the Universe were the same, it followed that there must be other planetary systems similar to our own, and that some of those planets would possess life.[20] However, not unnaturally given his interest in the details of spaceflight, Tsiolkovsky also firmly believed that mankind would construct habitats in the Solar System and then move out into space. His feelings were revealed in his famous phrase: "Earth is the cradle of intelligence, but it is impossible to live forever in the cradle." The monist in him impelled him to argue that if *we* expand into space, then all those *other* species must do the same. The logic is inescapable, and Tsiolkovsky was aware that this led to a paradox when maintaining both that mankind will expand into space *and* that the Universe is brimful with intelligent life. In 1933, long before Fermi asked his question, Tsiolkovsky pointed out that people deny the existence of ETCs because (i) if such civilizations existed, then their representatives would have visited Earth, and (ii) if such civilizations existed, then they would have given us some sign of their existence. Not only is this is a clear statement of the paradox, Tsiolkovsky offered a solution: he believed that advanced intelligences — "perfect heavenly beings" — consider mankind to be not yet ready for a visitation.[21]

Tsiolkovsky's technical works on rocketry and spaceflight were widely discussed, but the rest of his output was generally ignored in the Soviet era. An appreciation of his discussion of the paradox therefore came only recently. (Fermi's own contribution did not fare much better. In their influential 1966 book *Intelligent Life in the Universe*, Sagan and Shklovsky introduce a chapter with the quote "Where are they?"; they attribute it to Fermi, but they incorrectly state that it was uttered in 1943. In a later paper, Sagan says that Fermi's quote was "possibly apocryphal.")

In 1975, English engineer David Viewing clearly stated the dilemma. A quote from his paper encapsulates it nicely: "This, then, is the paradox: all our logic, all our anti-isocentrism, assures us that we are not unique — that they *must* be there. And yet we do not see them." Viewing acknowledges that Fermi was first to ask the important question — "Where are they?" — and that this question leads to a paradox. To my knowledge, then, this paper is the first that refers directly to the Fermi paradox.[22]

However, it was a 1975 paper by Michael Hart in the *Quarterly Journal of the Royal Astronomical Society* that sparked an explosion of interest in the paradox.[23] Hart demanded an explanation for one key fact: there are no intelligent beings from outer space on Earth at the present time. He argued that there are four categories of explanation for this fact. First, "physical explanations," which are based on some difficulty that makes space travel unfeasible. Second, "sociological explanations," which in essence suppose that extraterrestrials have chosen not to visit Earth. Third, "temporal explanations," which suggest that ETCs have not had time to reach us. Fourth, there are explanations arguing that perhaps they *have* been on Earth, but we do not see them now. These categories were meant to exhaust the possibilities. Hart then forcefully showed how none of these four categories provide a convincing account of the key fact, which led him to offer his own explanation: *we are the first civilization in our Galaxy*.

Hart's paper led to a vigorous debate, much of it appearing in the pages of the *Quarterly Journal*. It was a debate that anyone could enter — one of the earliest contributions came from the House of Lords at Westminster![24] Perhaps the most controversial offering came from Frank Tipler, in a paper with the uncompromising title "Extraterrestrial Intelligent Beings Do Not Exist." Tipler reasoned that advanced ETCs could use self-replicating probes to explore or colonize the Galaxy cheaply and in a relatively short time. The abstract to Tipler's paper sums it up: "It is argued that if extraterrestrial intelligent beings exist, then their spaceships must already be present in our Solar System."[25] Tipler contended that the SETI program had no chance of success, and was therefore a waste of time and money. His argument poured oil on the fires of the debate and led to a further round of argument. The coolest and best summary of the arguments came from David Brin, who called the paradox the "great silence."[26]

In 1979, Ben Zuckerman and Michael Hart organized a conference to discuss the Fermi paradox. The proceedings were published in book form,[27] and although the volume contains a variety of views it is difficult to read it without concluding that ETCs have the means, motive and opportunity to colonize the Galaxy. The means: interstellar travel seems to be possible, if not easy. The motive: Zuckerman showed how some ETCs would be forced into interstellar travel by the death of their star, and in any case it seems a wise idea for a species to expand into space to guard against the possibility of planetary disaster. The opportunity: the Galaxy is 13 billion years old, but colonization can take place over a period of only a few million years. Yet we do not see them. If this were a murder mystery, we would have a suspect but no body.

Not everyone was struck by the force of the argument. A recent book by the mathematician Amir Aczel makes the case for the probability of extraterrestrial life being 1.[28] The physicist Lee Smolin wrote that "the argument for the non-existence of intelligent life is one of the most curious I have ever encountered; it seems a bit like a ten-year-old child deciding that sex is a myth because he has yet to encounter it."[29] The late Stephen Jay Gould, referring to Tipler's contention that ETCs would deploy probe technology to colonize the Galaxy, wrote that "I must confess that I simply don't know how to react to such arguments. I have enough trouble predicting the plans and reactions of people closest to me. I am usually baffled by the thoughts and accomplishments of humans in different cultures. I'll be damned if I can state with certainty what some extraterrestrial source of intelligence might do."[30]

It is easy to sympathize with this outlook. When considering the type of reasoning employed with the Fermi paradox, I cannot help but think of the old joke about the engineer and the economist who are walking down a street. The engineer spots a banknote lying on the pavement, points to it, and says, "Look! There's a hundred-dollar bill on the pavement." The economist walks on, not bothering to look down. "You must be wrong," he says. "If there were money there, someone would already have picked it up."[31] In science it is important to observe and experiment; we cannot know what is out there unless we look. All the theorizing in the world achieves nothing unless it passes the test of experiment.[32]

Nevertheless, surely Hart's key fact *does* require an explanation. We have been searching for ETCs for more than 40 years. And the continuing silence, despite intensive searches, is beginning to worry even some of the most enthusiastic proponents of SETI. We observe a natural universe when we could so easily observe an artificial universe. Why? Where *is* everybody? Fermi's question still demands an answer.

FIGURE 10   *Enrico Fermi, sailing off the island of Elba. The photograph was taken shortly before his death.*

3

They Are
Here

The simplest resolution of the Fermi paradox is that "they" are already here; or, at least, "they" were here in the past. Of the three classes of solution to the paradox, this is by far the most popular among the general public: the notion that UFOs are alien spacecraft is accepted by many people, while the idea that ancient structures were built by extraterrestrials rather than by people is believed almost as widely. Scientists are much more skeptical, mainly because of the poor quality of the supporting evidence. Nevertheless, it is worth considering these ideas seriously as potential resolutions of the paradox. Indeed, some serious scientists would argue that, until we have explored our neighborhood much more thoroughly and can definitely rule out the presence of alien artifacts, there really *is* no Fermi paradox.

I interpret the title of this chapter rather loosely: I consider "here" to be not just Earth but the whole Solar System — and even, in the final two sections of this chapter, our entire Universe. To begin, though, I discuss the very first suggested resolution of the paradox. It was given to Fermi soon after he posed his question.

Chapter 3

SOLUTION 1: THEY ARE HERE AND THEY CALL THEMSELVES HUNGARIANS

... the cleverest man I ever knew, without exception.
Jacob Bronowski on John von Neumann in *The Ascent of Man*

The first answer to Fermi's question came almost immediately. Leo Szilard, one of Fermi's regular lunchtime companions at Los Alamos, joked: "They are among us and they call themselves Hungarians."

There was a whimsical story, often told within the Theoretical Division at Los Alamos, that Hungarian people are Martians.[33] Millions of years ago, so the story went, the Martians left their own planet and traveled to Earth, landing in what is now Hungary. At that time the European tribes were barbarians, so the Martians had to pass themselves off as human — if the barbarians suspected aliens were in their midst, then blood (or rather the Martian equivalent) would be shed. Except for three traits, the Martians successfully hid their evolutionary differences. The first trait was wanderlust: this found its outlet in the Hungarian gypsy. The second trait was language: Hungarian is unrelated to any of the Indo-European languages spoken in the neighboring countries of Austria, Croatia, Romania, Serbia, Slovakia, Slovenia and Ukraine. The third trait was intelligence: their brainpower was beyond that of mere humans.

Unfortunately for the theory, many peoples have exhibited wanderlust at some point in their history; and the Hungarian language is hardly unique, related as it is to Finnish, Estonian and some languages spoken in Russia. But that third trait was in evidence at Los Alamos: Fermi's lunchtime companions regularly included not only Szilard himself, but also Eugene Wigner, Edward Teller and John von Neumann. All four had been born in Budapest within ten years of each other. Another Hungarian at Los Alamos, Theodore von Kármán, was also a native of Budapest, but had been born slightly before the others. These "Martians" certainly constituted a formidable array of intellect. The physicist Szilard made contributions in several fields. Teller went on to be the prime mover behind the development of thermonuclear weapons. Wigner won the 1963 Nobel Prize in physics for his work in quantum theory. The engineer von Kármán performed early work in rocketry and the theory of supersonic drag, and his research led to the design of the first aircraft to break the sound barrier.

Easily the most brilliant of the Martians, though, was von Neumann. John von Neumann, whom we shall meet again later in the book, was one of the outstanding mathematicians of the 20th century. He developed the discipline of game theory, made fundamental contributions to quantum theory, ergodic theory, set theory, statistics and numerical analysis, and gained fame when he helped develop the first flexible stored-program digital computer. Toward the end of his career he was a consultant to big business and the military, allotting time to various projects as if his brain were a time-share mainframe computer. His ability to calculate in his head the answers to mathematical problems was legendary — he routinely beat Fermi whenever the pair had a calculating contest — and his near-photographic memory just added to an aura of unearthly intelligence. He possessed other talents that chimed nicely with the "Hungarians are aliens" story. "Good-Time Johnny" absorbed large amounts of alcohol at Princeton parties with seemingly no detriment to his mental faculties. He was involved in road traffic accidents at alarming rate — one junction in Princeton was known as "von Neumann Corner" after all the accidents he caused there — yet he always walked away unscathed. (The natural conclusion is that alcohol affected his driving, but there is no evidence that this was the case; he seems just to have been a bad driver.)

But even the "cleverest man in the world" sometimes got it wrong. Although he played a pivotal role in the development of the digital computer, and has thus affected our lives in a way that few mathematicians have done, von Neumann apparently thought that computers would always be huge devices, useful only for building thermonuclear bombs and controlling the weather. He failed completely to foresee a day when computers would be embedded in everything from the toaster to the tape deck. Surely a real Martian would have known better.

SOLUTION 2:   THEY ARE HERE AND ARE MEDDLING IN HUMAN AFFAIRS

What one man can fantasize, another man will believe.
William K. Hartmann

Shakespeare has Juliet ask: "What's in a name?" In certain situations the answer is: everything. For example, for thousands of years people have seen strange lights in the sky.[34] No great attention was paid to the phenomenon until the lights acquired a catchy name. Call them "flying saucers" and suddenly everyone is interested.

We can date the precise moment when a person first saw a "flying saucer." On 24 June 1947, Kenneth Arnold was flying his private plane over the Cascade Mountain range in Washington State. From his cockpit he saw several airborne objects; when he landed he reported his sighting, describing the objects as skipping "like saucers across a pond." The name stuck. The press was hungry for gossip about these "flying saucers," and the term found resonance with an American public nervously entering the

Cold War. Many people took it for granted that the flying saucers were crewed by aliens — either Russians or extraterrestrials.[35]

If flying saucers are real, if they are indeed spacecraft crewed by aliens, then the Fermi paradox is instantly resolved. Of all the proposed resolutions of the paradox, this one has most support with members of the public. As surveys consistently show, a majority of Americans believe flying saucers are visiting Earth right now; the proportion of Europeans holding that belief is smaller, but is still significant. Many people even believe that a flying saucer crashed at Roswell, New Mexico, in late June/early July of 1947 (suspiciously close to the time of Arnold's sighting), and that the US military recovered alien bodies from the wreckage. Nevertheless, science is not a democratic process. Hypotheses are not proven right or wrong through a ballot. No matter how many people believe in the truth of a particular hypothesis, scientists will accept the hypothesis (and then just provisionally) only if it explains many facts with a minimum of assumptions, if it can withstand vigorous criticism, and if it does not run counter to what is already known. So the question is: how well does the hypothesis that flying saucers are evidence of ETCs stand up?

*　　　*　　　*

Before discussing this, it is best to agree to use the neutral term "unidentified flying object," or UFO, when examining claims about strange lights or objects in the sky. The term was coined by Edward Ruppelt, who undertook an investigation of UFOs for the USAF.[36] Unfortunately, the terms "UFO" and "flying saucer" are often used interchangeably. But if used correctly, a UFO is just that: an aerial phenomenon that is *unidentified*. Everything we see in the atmosphere is either a UFO or an IFO (an identified flying object). Only upon investigation can a UFO become an IFO; an IFO *might* turn out to be a flying saucer — but only after careful scrutiny can we make that determination.

Under this definition, it is undeniable that UFOs exist. Indeed, it is tempting to say that if you have not seen a UFO, then you have not been looking hard enough! The sky is host to a myriad of interesting phenomena, both natural and artificial. Nevertheless, upon even a cursory examination most UFOs are explicable; they become IFOs. People often mistake Venus for an artifact; aircraft can create unusual visual effects; each day, 4000 tons of extraterrestrial rock and dust burn up in the Earth's atmosphere and produce the occasional light show; and so on. Some UFOs become IFOs only after a thorough and detailed investigation. (For example, the novaya zemlya, fata morgana and fata bromosa mirages have fooled people for hundreds of years. They are caused by relatively rare atmospheric conditions; perhaps the same mechanism can explain some UFOs?

Maybe some of those strange lights in the sky are the beams of car head-lights refracted through abnormal air conditions?) A few UFOs might be the result of accidents (one mysterious light turned out to be the result of a golfball thrown onto a bonfire — who knows what other effects every-day events might produce?). The explanation of some UFOs might even require advances in science (the phenomenon of ball lightning, for exam-ple, is poorly understood and not well researched — ironically for the same reasons that many scientists feel uncomfortable with the idea of UFOs). Fi-nally, many UFOs turn out to be the result of deliberate hoaxes.

Upon investigation, then, most UFOs become IFOs. But each year has a tiny residue of cases in which no rational account is forthcoming. We should not find this surprising. After all, as the noted skeptic Robert Sheaffer points out, po-lice do not achieve a 100% so-lution rate for murders.[37] But many people find this unac-ceptable when discussing UFOs; they want an explanation for *all*

FIGURE 11   One of the most famous of all photographs of a "flying saucer." It was taken on 11 May 1950 by Paul Trent at his farm at McMinnville, Oregon.

sightings. How can we explain these UFOs? There are two cases to consider: sightings of lights in the sky, and sightings of — perhaps even encounters with — aliens or alien technology.

If a reported UFO was simply a light in the sky, then one could argue that, no matter how strange it appeared to be, we do not *have* to explain it. Life is too short for scientists to explain every instance of every phe-nomenon. A scientist no more has to explain the detailed circumstances that produced a particular light in the sky than he has to explain the shape of the strange dragon-like cloud formation I saw this morning as I was walking to work. There are more important things to study.

But what if an explanation is *demanded*? My feeling is that we need no new hypothesis to explain the anomalous sightings: the reasons that ac-count for most UFOs would account for *all* UFOs if we were clever enough (and if we had enough time) to carry through the investigations. Sheaf-fer highlights the interesting finding that the percentage of "inexplicable" UFOs does not vary much within the overall number of sightings. In other words, whether it is a busy year or a quiet year for UFO sightings, the IFO/UFO ratio is about the same. This is not at all what one would expect if the "inexplicable" UFO sightings represented alien craft. The simplest explanation of this finding is that, in Sheaffer's words, "the apparently unexplainable residue is due to the essentially random nature of gross mis-perception and misreporting."

None of this proves that we are *not* receiving visits from ETCs. (Nor does it prove that when we see UFOs we are not watching manifestations of ghosts, fairy craft or the sporadic intersection of higher-dimensional be-ings with our own spacetime.) But neither does the observation of UFOs prove that we *are* receiving visits. The cast-iron, unimpeachable sightings of lights in the sky are just that: sightings of lights in the sky. The existence of unidentified aerial phenomena simply does not provide any evidence for the existence of extraterrestrial visitations.

What if the reported UFO was something *more* than a light in the sky? How can we explain the "close-encounter" sightings? Unfortunately, the *interesting* sightings, the events that would prove the flying saucer hypoth-esis true, are all in some way problematic.

There are claims, for example, of individuals being abducted by aliens, being probed, being forced to have sex. However, no matter how plausi-ble you find these stories (I freely admit to bias; I find the stories utterly implausible, since the chances of totally separate evolutionary lines pro-ducing organisms morphologically similar enough to have sex are surely infinitesimal), the *evidence* required to support such claims is non-existent.

There are reports that alien craft have crashed; the Roswell incident, mentioned above, is well known. But once again, whether or not you find it likely that a craft can successfully travel interstellar distances yet fail to negotiate a planetary atmosphere, the *evidence* in favor of such reports is shoddy. An item of advanced equipment or a sample of an unknown alloy would prove the case; instead, we are given a video of an autopsy of one of the "aliens" from the crashed Roswell vehicle — a video that was, of course, a (profitable) hoax.

There are claims that alien craft have landed in various countries. In England, for example, UFOs have been blamed for the crop circle phe-nomenon. At least some, and maybe all, of the crop circles are man-made. In a recent case, a self-confessed maker of crop circles got into trouble with the law. He made a 7-pointed shape after hearing an "expert" claim that elaborately designed crop circles were impossible for man to make. (Crop circles actually have a variety of shapes; there are crop triangles, crop hexagons, even crop fractals.) Complex designs had been documented, so this was proof — according to the expert — that at least some crop circles were extraterrestrial in origin. The crop circle maker, armed only with some planks, bamboo poles and a torch, proceeded to create his 7-point shape over three nights in a field of ripening wheat. Personally, I admire his devo-tion to rationality, but the farmer was not impressed; neither was the judge, who issued a £100 fine for criminal damage. (And my guess is that, despite the demonstration, the expert is still of the opinion that crop circles are the landing marks of flying saucers.) In these situations, surely we should use Occam's razor, one formulation of which is that explanations of unknown phenomena should first be sought in terms of known quantities.[38] We can explain crop circles, cattle mutilations and other fringe phenomena in terms of known quantities. We do not need the flying saucer hypothesis to explain them.

Whenever an extraordinary claim is made for flying saucers, no extraordinary evidence is presented to support the claim. Instead, we get lies, evasions and hoaxes. The flying saucer hypothesis may be the most popular explanation of the Fermi paradox, but surely there are better explanations.

*   *   *

Incidentally, I should state here that I have seen a UFO, and it remains one of my most vivid memories. While playing soccer in the street as a child — this was before the increasing number of cars stopped children playing in the street — I looked up and saw a pure white circle about the size of the full moon. Protuberances on either side of the circle made it look rather like Saturn showing its rings edge-on. Whatever it was, it seemed to hover for a few seconds before moving off at tremendous speed. I was with a friend, who also saw it and remembers it still. Interestingly, we differ in our recollections: I remember it shooting away to our left as we watched; my friend says that it moved away to our right. (People are poor observers, and I know from experience that I am a *very* poor observer. But I am adamant that it moved to the left!) We definitely saw *something* in the sky that day and I have no idea what. But no, it was not a flying saucer. It was just a light in the sky.

SOLUTION 3: THEY WERE HERE AND LEFT EVIDENCE OF THEIR PRESENCE

> *Tell them that I came, and no one answered.*
> Walter de la Mare, *The Listeners*

The evidence that ETCs are currently visiting Earth is not compelling. But maybe they visited Earth, or at least our Solar System, some time in the past — perhaps long ago, at a stage in human development when no one could recognize them for what they were. Is there any evidence for this? Let us work through the Solar System, beginning at home.

Earth

The famous Tunguska explosion of 1908 — an event that felled acres of trees across the Siberian taiga — was long thought to be the result of an asteroid strike. Researchers, however, found none of the debris that one would expect from such an impact. It was a mystery. Once the immense power of nuclear explosions became apparent, soon after World War II, the notion circulated that the Tunguska event had been a nuclear blast — the impact of an alien nuclear-powered spacecraft that had crashed. The idea was taken semi-seriously, and there was a simple means of testing it: go to Tunguska and search for traces of radioactivity. This was done, and scientists found no traces of radioactivity that could have come from a nuclear engine. (They also ruled out an antimatter engine.) We now know the Tunguska event was probably the result of a stony meteoroid that exploded in the atmosphere (although the evidence is still not conclusive, and several scientists believe Tunguska was hit by a comet). There have been similar events in the past, and they have a similar explanation: meteorite impact. There is no need to invoke the hypothesis of a downed spaceship. If a spaceship ever did crash-land in the past, we have not found the evidence (Roswell notwithstanding).

In the 1970s, Erich von Däniken became famous for a series of books in which he claimed that extraterrestrial visitors built many of the enigmatic structures dotted around the world — Stonehenge, the lines on the Nazca Plain in Peru, the Easter Island statues, and so on.[39] None of the books contained proof to back up his claims. Nevertheless, his large reading public supported him during his lengthy spell in prison for fraud; they supported him after his claims were painstakingly and thoroughly debunked; only when they became bored and taste and style moved on did they drop him. Now, like several pop groups from that era, von Däniken and his ideas are back in fashion even though, in the thirty-odd years since the books were first published, no proof has been produced to support his speculations — something that von Däniken himself cheerfully admits and seems to find irrelevant. Since the supporters of von Däniken are unlikely to be swayed by rational argument, we may as well move on — and accept that there is no evidence that members of an ETC have ever been on Earth. (This, of course, is not to say that they definitely have *not* been here. If they visited Earth 1 billion years ago, say, who knows what signs — if any — would remain of their visit? But in the absence of any evidence to the contrary, we may as well assume that Earth has been untouched.)

Moon

Until fairly recently, some people claimed to see evidence for ETCs on the Moon. In 1953, for example, the astronomer Percy Wilkins discovered what appeared to be an artificial structure — a bridge.[40] However, other astronomers could not see the structure through more powerful telescopes and decided, quite reasonably, that the bridge was a trick of the light. This did not dampen the enthusiasm of those who believed in the Moon as an abode of alien life. Enthusiasts pointed out that the Moon shows only one side to Earth (to be precise, due to the phenomenon of libration we see only 59% of the Moon's surface). If we never see 41% of the lunar surface, who knows what might be hiding on the far side of the Moon? It was not until the late 1970s, well after the many landers and orbiters had mapped the entire surface of the Moon, that "life" enthusiasts finally stopped promoting the idea of bridges and other artifacts. (At least, I *think* they have stopped promoting the idea.)

Earth–Moon Lagrangian Points

As we shall see later (page 79), one can argue that an ETC wishing to explore our Solar System would send small unmanned (unaliened?) probes rather than a fleet of crewed spacecraft. Where might we find such probes? There are three cases to consider. First, the probes could be programmed to attract our attention. Since we see no evidence for beacons, it is safe to assume that such probes are not here. Second, the probes could be programmed to hide from us. The Solar System is a large place, and there are plenty of places where they could hide. Since we are unlikely ever to find such probes, we need not spend time discussing the best strategy to observe them. Third, an ETC might send probes and not care whether humans observe them. If that is the case, where might we find them?[41]

We can reasonably argue that of all the planets in the Solar System, ours is most worthy of study. Earth is an interesting planet for a variety of reasons — most importantly it is, as far as we know, the only planet to harbor life. So probes would most likely be programmed to investigate Earth. (This argument of course reeks of anthropocentricism. Who knows what an alien mind might want to investigate? Who knows what technology it might employ? But such logic is all we have, so we lose nothing if we continue the argument and see where it leads us.) The surface of the Earth would be a poor site for long-term studies of our planet. It would make more sense to view the entire planet from space, where solar energy is more readily available, and where there is no need for the probe to protect itself against the effects of the Earth's geological activity.[42]

Chapter 3

Several types of orbit are suitable for long-term parking of observational probes, but perhaps the best known are the *Lagrangian points* L4 and L5.[43] If a small mass is near two large orbiting masses, then there are five points at which the small mass can orbit at a fixed distance from the larger masses. These five Lagrangian points mark the positions where the gravitational pull of the two larger masses exactly balances the centripetal force required to rotate with them. At first glance, then, there are five points where ETCs might place a small probe in the hope that it maintains a fixed distance from Earth and Moon. However, three of the Lagrangian points — L1, L2 and L3 — are unsuitable because they are unstable: nudge the small mass and it will move away from the L point. But L4 and L5 are stable: nudge the small mass and it will return to the L point. (To be precise, L4 and L5 are stable only if the most massive of the three bodies is at least 24.96 times as massive as the intermediate body. This condition is satisfied in the Sun–Earth system, since the Sun is much more massive than Earth. It is also satisfied in the Earth–Moon system, since Earth is 81 times as massive as the Moon. The Sun's gravitational influence tends to destabilize the L4 and L5 points of the Earth–Moon system; however, it smears the stable *points* into *volumes* of space in which stable orbits exist.)

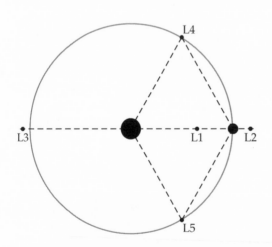

FIGURE 12   *The five Lagrangian points are places in the vicinity of two orbiting masses where a third smaller body can maintain a fixed distance from the larger masses. The points L1, L2 and L3, which lie on a line connecting the two large masses, are unstable: after a perturbation, the small body will move away from the Lagrangian point. Under certain circumstances, the points L4 and L5 are stable: after a perturbation, the small body will return to the Lagrangian point. The L4 and L5 points are stable for the Earth–Moon system, so they are a good place to park probes for long-term study of the Earth.*

NASA is already using the Lagrangian points of the Sun–Earth system as parking places for its satellites. The L1 point is home to SOHO (Solar and Heliospheric Observatory); from the L1 vantage point, SOHO has an uninterrupted view of the Sun. The L2 point is home to MAP (Microwave Anisotropy Probe); from there, MAP will study wrinkles in the cosmic microwave background and uncover information about the Big Bang. If NASA finds it convenient to park satellites at L points, then perhaps ETCs would do so too. Perhaps we might find probes at Lagrangian points in the Earth–Moon system? Well, at least one dedicated search has been made. Furthermore, astronomers have already studied the L4 and L5 points of the Earth–Moon system, since the points are interesting from a general astronomical viewpoint — material will tend to accumulate there. (The Trojan asteroids Agamemnon, Achilles and Hector, for example, orbit in the L4 and L5 points of the Sun–Jupiter system.) However, in neither the dedicated search nor the general scans was any evidence of probes found.

Increasingly, other near-Earth orbits are being scanned — this time by astronomers searching for potentially lethal asteroids. As a by-product of this research we might hope to find artifacts; so far, though, none have been found. Probes would give off heat, but no anomalous infrared signals have been observed; probes might be expected to transmit messages back to their creators, but no such transmissions have been detected.

Some people have claimed that long-delayed radio echoes (LDEs) are transmissions from ETC probes. The LDE phenomenon — radio echoes that appear between 3 and 15 seconds after transmission of the signal — has been observed since the dawn of radio, and it remains somewhat mysterious. Radio echoes from the Moon are common, but the echo appears 2.7 seconds after transmission of the main signal — this being the time it takes light to travel to the Moon and back. Echoes from Venus, the nearest planet, can only appear 4 minutes after the main signal. So neither the Moon nor Venus can be the cause of LDEs. One explanation is that they are radio returns from ETC probes that are beyond the distance of the Moon. A more prosaic explanation is that they are a natural phenomenon caused by plasma and dust in the Earth's upper atmosphere.[44]

Although the search for probes is not complete — indeed, the search has hardly begun, as Earth could be bathed in signals at certain frequencies and we would not necessarily know about them — all observations to date have given a negative result. (Interestingly, our telescopes *have* occasionally detected transmissions from a probe in the depths of our Solar System; but they are from the Pioneer spacecraft, not from an ETC craft.)

Mars

Mars has long been thought to be home to life,[45] but much of the fuss stemmed from a mistranslation.[46] Giovanni Schiaparelli, in a series of observations beginning in 1877, saw features on Mars that he called *canali* — an Italian word meaning "channels" or "canals." It is clear from his writings that Schiaparelli, when he named these features, thought *natural* processes had formed them. English-speaking astronomers, however, translated the word as "canals" — *artificial* structures connecting two bodies of water.

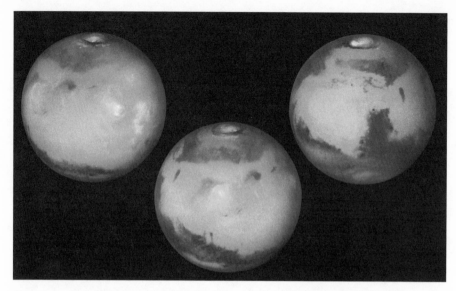

FIGURE 13   *Mars as photographed by the Hubble Space Telescope, when the red planet was at its closest to Earth.*

Percival Lowell also saw the surface features recorded by Schiaparelli, and he finally counted 437 of them.[47] However, Lowell failed to acknowledge he was working at the limits of observation; he did not realize evolution has primed the human visual system to look for familiar features in random patterns. He became convinced he saw artificially constructed linear canals, and he speculated that the canals supplied water from the polar caps to a desert world. The notion of canals was in the public consciousness anyway — the Suez Canal, a modern wonder of the world, had opened to navigation in 1869 — and the general public was gripped by the possibility that intelligent beings had constructed the Martian canals. Science fiction writers were quick to use it as a source of stories. It was a popular and romantic notion, and even as late as 1960 some maps of the planet showed oases and canals; and several astronomers continued to believe that seasonal changes in the Martian surface markings might be due to changing vegetation patterns.

FIGURE 14   *Percival Lowell.*

Meanwhile, in the early 1960s,[48] Shklovsky discussed a peculiarity in the orbit of Phobos, the larger of Mars' two moons, and offered an ingenious explanation.

The orbit of Phobos is decaying. The peculiarity was that, according to observations made by Bevan Sharpless in the 1940s, the *rate* of decay was difficult to explain. A number of mechanisms was suggested — the effect of a hypothetical large Martian magnetic field, tidal interaction with Mars, a possible solar influence — but none of them were feasible.

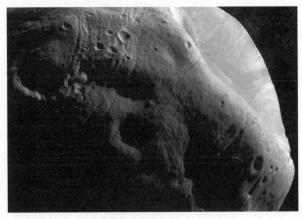

FIGURE 15   *Phobos, the larger of Mars' two moons, is a potato-shaped rock about 16 miles by 10 miles in size. It is almost certainly a captured asteroid.*

Neither was the obvious explanation that Phobos was passing through the thin outer regions of the Martian atmosphere, as drag would not affect a rock the size of Phobos to the extent observed by Sharpless. The audacious Shklovsky wondered whether Phobos were *hollow*. A hollow Phobos would be less massive than its size would suggest, so its orbit would be affected more by the Martian atmosphere. If Phobos really were hollow, then it could not be natural: Shklovsky thus suggested that the satellite was artificial — the product of a Martian civilization. (It was a suggestion more imaginative than anything in the books of von Däniken, yet it was based on the best available observational data.) Shklovsky thought the satellite would have been launched millions of years ago, but other scientists thought the launch could have been more recent. Frank Salisbury pointed out that the Martian moons were discovered in 1877 by Asaph Hall, who used a 26-inch telescope.[49]   Fifteen years earlier, when Heinrich d'Arrest trained a larger telescope on the red planet, the conditions for viewing Mars had been better. How could d'Arrest have missed the moons in 1862? Was it possible, Salisbury asked, that the moons were artificial satellites launched between 1862 and 1877?

The romantic notion of an advanced Martian civilization capable of building canals and launching satellites did not survive the 1960s. It was laid to rest when the early Mariner spacecraft flew by at close range, returning photographs that showed none of the canals seen by Lowell. The Viking

FIGURE 16   *The "face" on Mars. This low-resolution image contains many black dots, which are artifacts of the image-processing techniques employed by the Jet Propulsion Laboratory, and do not correspond to any Martian feature.*

landers of 1976 and the Pathfinder and Mars Global Surveyor missions of 1997 also failed to find canals. Similarly, the flyby missions showed that there is nothing artificial about Phobos. It is a small pockmarked piece of rock — almost certainly a captured asteroid. Furthermore, although its orbit is indeed decaying, recent measurements indicate that the rate of decay is only half that calculated by Sharpless. With this improved measurement, theorists can now explain the origin of the drag on Phobos: it is the result of tidal interaction with Mars. (Phobos draws closer to Mars by about 1 inch every year. The satellite will hit Mars some time within the next 40 million years, leaving a basin the size of Belgium. Although 40 million years is a short time on the astronomical scale, it is a long time on the human scale. A pity — it would be a spectacular event.)

The evidence from the various flyby, orbiting and lander missions almost killed the belief in an ancient Martian civilization. Almost, but not quite. In 1976, Viking photographed the Cydonia region on Mars, and NASA released the photographs soon afterward. Almost immediately, enthusiasts pointed out that one of the low-resolution photographs appeared to show a human face. You could make out an eye, a mouth, and a nostril (though the enthusiasts often failed to point out that the "nostril" was actually an artifact of the way the image had been processed, and did not

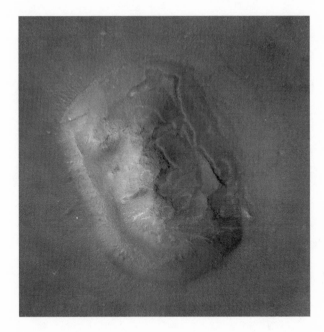

FIGURE 17   Is it a shield? Is it a footprint? Is it Chewbacca? A high-resolution image of the Cydonia region, this time taken by Mars Global Surveyor in 1998, shows no evidence for a face.

correspond to any physical structure on Mars). The face was large, roughly a square of 1 km, and seemingly carved out of stone. NASA scientists emphasized that this was a natural formation; the image was simply the result of sunlight falling on a hill one Martian afternoon. Others argued that the formation was an artificial structure; the stone "face" was proof that Mars was once home to an ancient civilization.

If you search through a large collection of random data long enough and hard enough, conveniently ignoring arrangements of the data that are of no interest and not defining beforehand what you are looking for, then eventually you will find something that seems remarkable. The surface of Mars covers 150 million km^2; it would be strange if one of those square kilometers did *not* vaguely resemble something familiar. Planetary scientists argued that the Martian "face" has as much significance as the patterns you see in the coals on a fire. It was another instance of an observer imposing meaning on a meaningless pattern.

Mars Global Surveyor revisited the Cydonia region and took a more detailed photograph. The evidence for the face, of course, evaporated. (It is only fair to point out that the illumination is different in the two photographs. Nevertheless, modern computer imaging techniques can retain the detail of the Global Surveyor photograph while simulating the feature in the same afternoon light that Viking saw. If I scrunch my eyes, then I can just about make out Chewbacca from *Star Wars* — but no human face.)[50]

41

Chapter 3

Asteroids

Michael Papagiannis argued that we must rule out the possibility that ETCs are in the Asteroid Belt before we can conclude that they are not here.[51] The Asteroid Belt would be an ideal place for ETCs to set up space colonies. They could mine the asteroids for natural resources, and they would have plentiful supplies of solar energy. Who knows — perhaps the fragmentation of the Asteroid Belt components is the result of large-scale mining projects by ETCs? If space colonies *were* in the Asteroid Belt, we would not necessarily know about them; craft that were, say, 1 km or less in size would be difficult to distinguish from natural asteroids.

On the other hand, if they really are in the Asteroid Belt, there are questions to ask. Why have we detected no leakage of electromagnetic radiation? Why have we not observed a single object that possesses an effective temperature higher than is justified by its distance from the Sun? And why, if they are there, have they remained silent for so long?

Outer Solar System

Beyond the asteroids we can see numerous "anomalies" — like the axial tilt of Uranus or the retrograde orbit of Triton — that could be taken as evidence for tampering by ETCs. David Stephenson, for example, suggested that Pluto's unusual orbit is the result of an astroengineering project.[52] These anomalies, however, can be explained more prosaically as the result of collisions and interactions that took place in the early history of the Solar System. There is simply no need to invoke other explanations.

*   *   *

When we begin to discuss the outer planets we also begin to realize just how big the Solar System is. There are 50 billion billion billion cubic miles of space within a sphere that encloses the orbit of Pluto; and the Solar System extends to the Oort Cloud of comets, *far* beyond Pluto. The chances of finding a small alien artifact by accident are essentially zero. Only if an artifact draws attention to itself — by signaling us, perhaps, or by being in a visible location — will we detect it. We therefore cannot rule out the possibility that observational probes were once in the Solar System nor, indeed, that they are still here. Some would argue that until we *can* rule out that possibility, there is no Fermi paradox.

We can say with confidence, however, that no evidence for alien artifacts has yet been uncovered.[53] If we do not observe them, why assume they might be there? (Besides, if probes *are* in the Solar System, we are still left with the problem of why they have left Earth alone.)

Back to Earth

Perhaps we are looking in entirely the wrong place. The discussion has revolved around alien artifacts — evidence of engineered objects. Perhaps an ETC has been here and left *information* rather than *things*?

An entertaining SF story from the 1950s suggested that the reason so many people dislike spiders is that the class Arachnida consists of alien creatures. They were carried here on some spacecraft, and then escaped; humans, instinctively recognizing the spiders' alien heritage, recoil from them. (As we shall see later [page 189], all life on this planet is related; however much you may dislike spiders, you share a large part of your DNA with them.) In the 1970s, some scientists finally caught up with SF writers and made the suggestion that biological material might carry a coded message from an ETC. In theory, this would be possible: after all, the whole point of DNA is that it encodes information.

A message encoded in DNA seems an unlikely communication channel. For one thing, the sender could convey a message only to a planet that possessed the same biochemistry. (In our case, the sender's biochemistry would have to be based on L-amino acids, have protein synthesis based on the same genetic code as ours, and so on.) Even if it were possible for the recipient to distinguish between a natural sequence and an artificial message, over time the content of the message might become garbled through random mutations. And the vagaries of evolution might erase the message altogether. Nevertheless, a few investigations have been performed to test the idea,[54] and analysis of certain types of viral DNA has found nothing resembling an artificial pattern. Now that biologists have sequenced the entire genome of several creatures, including man, more detailed searches could be performed for coded messages. Such searches must be low on the list of priorities for geneticists, but eventually someone will sift through the genome data looking for patterns. My guess is that patterns *will* be found, but they will have the same source as the Martian canals and the Cydonian face. Such patterns are evidence of intelligence — but at the observer's end of the telescope or microscope.

SOLUTION 4: THEY EXIST AND THEY ARE US — WE ARE ALL ALIENS!

I should have known what fruit would spring from such a seed.
Lord Byron, *Childe Harold*

In the previous section we considered the idea that ETCs might have encoded messages in the DNA of terrestrial organisms. Although this is a remote possibility, a broader version of the idea is, paradoxically, more plausible. With each breakthrough in the study of genetics it becomes increasingly apparent that all life on this planet is deeply related. Perhaps individual species are not alien, but we cannot discount the possibility that *every* species came from the same extraterrestrial source. Perhaps we are *all* aliens.

The idea that life originated elsewhere and was somehow transported to Earth is an old one. The notion of *panspermia* — literally "seeds everywhere" — probably dates back to Anaxagoras. Some of the best scientists of the 19th century discussed various forms of panspermia, but it was a book by Arrhenius in 1908 that popularized the notion. Arrhenius supposed that the Universe is full of living spores that are driven through space by the pressure of starlight. Such spores fell on the early Earth, flourished, and evolved into the life we see today.[55]

As we shall discuss in more detail later (see page 189), one of the deep mysteries of the origin of life is the indecent haste with which it arose on Earth. There scarcely seems enough time for random physical and chemical processes to generate life from lumps of inanimate matter. The panspermia idea is attractive, since it removes the problem of timescales: life dropped "ready-made" onto Earth. Nevertheless, the Arrhenius hypothesis quickly fell from favor. One reason why the idea was shelved was the difficulty of imagining spores hardy enough to withstand the rigors of an aeons-long journey through space; in particular, radiation would surely prove deadly to spores. Another reason was that it merely removed the problem of the ultimate origin of life from Earth to somewhere in space (although of course it would be nice to know where life originated, if only to settle a fact of history).

The idea that there may be microbial life out in space did not entirely disappear. For example, Hoyle and Wickramasinghe championed the idea that microbes travel to Earth on comets, causing occasional mass outbreaks of disease.[56] The claim was lent some credence by the discovery that bacteria traveled to the Moon on unmanned lunar landers, and were still alive when brought back to Earth by Apollo astronauts. Clearly some microorganisms can survive the harsh environment of space, if only for a few years.

Furthermore, the much-hyped announcement in 1996 that the Martian meteorite ALH84001 might contain bacterial microfossils led some scientists to suggest that life began on Mars. Microbes subsequently traveled to Earth inside meteorites, which would protect them from the harsh environment of space. It is an attractive suggestion: conditions on early Mars may well have been more conducive for the emergence of life than those on early Earth. However, skepticism is in order. Recent work suggests that the so-called microfossils may be artifacts of the procedures used to view the rock at extreme magnification. The ALH84001 affair is perhaps yet another example where Mars has led scientists astray, causing those who are working at the very limits of observation to see patterns that are not there.

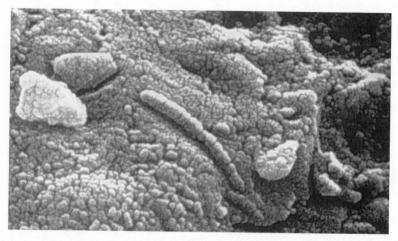

FIGURE 18   *When viewed under extreme magnification, the rock ALH84001 — a meteorite from Mars — contains these strange worm-like structures. Are these structures microfossils of Martian bacteria, or artifacts of the techniques used to view the rock at such high magnification?*

Although panspermia is not in the mainstream of biological thought, it is a possibility that has certainly not been ruled out. If it turns out to be true, then the chances of life being a frequent occurrence in the Universe are greatly increased (though it does not necessarily say anything about the existence or otherwise of *intelligent* life and ETCs). In 1973, however, Crick and Orgel published the idea of *directed* panspermia:[57] panspermia plus intelligence, as Dyson put it. Crick and Orgel felt that the chance of viable microorganisms landing on Earth after an interstellar journey measured in light years was small. But *deliberate* seeding is different. Directed panspermia is the suggestion that an ancient ETC may have deliberately aimed spores toward planets with conditions favorable to the survival of life. Maybe primitive life did not arrive here haphazardly inside a meteorite; maybe it was *sent* here via a probe. (Why would an ETC seed planets in this way? Perhaps they were preparing planets for subsequent colonization, but somehow failed to get round to colonizing Earth. Perhaps they were performing grand astrobiological experiments. Perhaps they faced a global catastrophe, and wanted to ensure the survival of their genetic material. Who can tell?)

It is difficult to know how to test the hypothesis of directed panspermia. Billions of years after the event, how can we distinguish between primitive life emerging from the primordial ooze, primitive life arriving inside a meteorite, or primitive life arriving by space probe? (In their paper, Crick and Orgel argued that directed panspermia could resolve certain puzzles. For example, why is there only one genetic code on Earth? A universal code follows naturally if all life on Earth represents a clone derived from a single set of microorganisms. Another example relates to the dependence of many enzymes on molybdenum. This metal is rather uncommon — it ranks 56th in order of abundance of the elements in the crust of the Earth — and yet it plays an important biochemical role. This slightly odd state of affairs would be less surprising if life on Earth derived from a system in which molybdenum was much more abundant. Of course biochemists have more orthodox answers to these puzzles, and so the evidence in favor of directed panspermia is weak.) If biologists develop a convincing theory of how life originated naturally from the materials available on the primordial Earth, then panspermia — directed or otherwise — would be unnecessary. Or Crick and Orgel might some day be proved true: we may meet the ETC that seeded our part of the Galaxy with spores. Until it is shown to be true or false, the hypothesis of directed panspermia remains on the table as a possible resolution of the Fermi paradox: ETCs exist because we sprang from their seeds. Where are they? They are here, because *we* are aliens.

SOLUTION 5: THE ZOO SCENARIO

> *Someone told me it's all happening at the zoo.*
> *I do believe it, I do believe it's true.*
> Paul Simon

In 1973, John Ball proposed the zoo scenario as a means of resolving the Fermi paradox.[58] (In fact, Ball called it the "zoo hypothesis"; variants of the idea, some of which are described below, also call themselves "hypotheses," and they appear as such in the literature. I prefer to call them scenarios, because in science an hypothesis usually implies a speculation framed in such a way that it can be tested. As we shall see, Ball's speculation cannot be tested. This is not to say the zoo scenario is untrue or is somehow more unlikely than other explanations. Indeed, we shall meet ideas that appear far more wild and improbable than Ball's speculation; but they merit the term "hypothesis" because they present testable predictions.)

Ball argued that ETCs are ubiquitous; many technological civilizations will stagnate or face destruction (from within or without), but some will develop their level of technology over time. Arguing in analogy with terrestrial civilizations, he reasoned that we need only consider the most technologically advanced civilizations. The advanced ETCs will, in some sense, be in control of the Universe; the less advanced will be destroyed, tamed or assimilated. The important question becomes: how will highly developed ETCs choose to exert their power? Arguing in analogy with how mankind exerts its power over the natural world, wherein we set aside wilderness areas, wildlife sanctuaries and zoos so that other species can develop naturally, Ball speculated that Earth is in a wilderness area set aside for us by ETCs. The reason there seems to be no interaction between them and us is that they do not want to be found, and they have the technological ability to ensure that we do not find them. The zoo scenario suggested that advanced ETCs are simply observing us. (Variants on the idea were less appealing; the laboratory scenario would have us as the subjects of laboratory experiments.)

This general idea has a long history in science fiction, predating Ball's publication. For example, *Star Trek* had the "Prime Directive," which stated that the Federation should not interfere with the natural development of a planet. (The Directive was more honored in the breach than the observance, of course, since the writers had to generate plots.) And before that, an established trope of the *Astounding* of the 1950s (under the strong but quixotic editorship of John Campbell, *Astounding* was the leading SF magazine of the day[59]) was of Earth under quarantine — either because ETCs were protecting us or, more commonly, because mankind was a threat to them. One could also argue that Tsiolkovsky's solution to the paradox — that ETCs have set Earth aside in order to let mankind evolve to a state of perfection — contains the seeds of the zoo scenario.

Believers in flying saucers tend to favor the zoo scenario as if it legitimizes their belief. Yet the zoo scenario specifically predicts that we should *not* see flying saucers or any other manifestation of superior technology. If flying saucers are spacecraft then the zoo scenario is wrong. (James Deardorff proposed a variant of Ball's idea, known as the leaky embargo scenario, which is compatible with observations of flying saucers. The idea is that advanced and benevolent ETCs have put in place an embargo on official contact with mankind. But the embargo is not total: aliens contact those citizens whose stories are unlikely to be credible to scientists and the government. The aliens want to slowly prepare us for the shock that might come later when they reveal themselves. Deardorff's proposal is so unscientific — though again not necessarily untrue — that it probably does not merit even the term "scenario."[60])

The zoo scenario has been attacked on several grounds. A major drawback is that it leads us nowhere; it is not a testable hypothesis. A good hypothesis generates ideas for observations that might confirm or falsify it, and in doing so generates new hypotheses. It is difficult to think of any observation that could test the validity of the speculation. Its one prediction is that we will not find ETCs, but the failure to find them hardly confirms the initial statement. There is something unsatisfying about an approach in which, no matter how hard we look, no matter how thoroughly we search, the absence of ETCs is explained simply by saying they do not want us to see them. (I can explain the lack of observational evidence for fairies at the bottom of my garden by saying they become invisible whenever people look their way. Irrespective of whether fairies exist, this is a poor sort of explanation from a scientific standpoint.)

Another criticism is that it is anthropocentric. Why should an ETC have any interest at all in a backward species like us? (Assuming, of course, that it is *us* they are interested in and not dolphins or monkeys or bees.) However, since we have absolutely no conception of what alien minds might find diverting, we cannot rule out the possibility that Earth — for whatever reason — has been set aside as the Galactic equivalent of a national park. A more serious weakness is that the zoo scenario fails to explain why Earth was not colonized long before complex life-forms appeared. Perhaps the scenario describes the reaction of ETCs to the discovery of intelligent life on Earth, but would the reaction be the same if all they found were primitive single-celled organisms?

A further criticism is that it takes only *one* ETC to break the embargo, and just one immature civilization to poke its fingers through the bars of the cage, for us to see them here on Earth. Furthermore, it fails to explain why we do not observe *any* evidence of them out there in the Galaxy. Ball proposes that advanced, intelligent life is ubiquitous. So where are their engineering projects? Where are their communications? It is one thing for them to keep Earth free from development, but quite another for them to stop all activity on our account.

Finally, it suffers in a way common to all solutions to the Fermi paradox that depend upon the motivations of alien intelligences. It supposes that *all* ETCs at *all* times behave in the same way with regard to us.

An expanded version of the idea, known as the interdict scenario, attempts to generalize Ball's idea and address some of the weaknesses.

SOLUTION 6: THE INTERDICT SCENARIO

Ever absent, ever near.
Francis Kazinczy, *Separation*

In 1987, Martyn Fogg proposed the interdict scenario — an expanded form of the zoo scenario that provides reasons why *all* life-bearing planets, not just Earth, are off limits.[61]

Fogg first presented the results of a simple model of the origin, expansion and interaction of early Galactic civilizations. Like many authors before him he found that, using seemingly plausible values for model parameters, the Galaxy fills relatively quickly with intelligent species. Depending upon the parameters, either a few species dominate with large "empires" or there are many different smaller "empires." The conclusion of Fogg's model is that, whatever the value of the parameters, ETCs would colonize the Galaxy even before our Solar System forms.

Fogg argues that once the colonization phase is over and nearly every star supports intelligent life-forms, the Galaxy enters a new "steady-state" era. The expansionist urge withers, and the problems of aggression, territoriality and population growth are solved. The distribution of intelligence becomes increasingly well-mixed and homogeneous, and the steady-state era becomes an age of communication. According to the model, we are billions of years into this (wonderful sounding) era.

If Fogg's scenario is true, then Earth is located within a sphere of influence of one or more advanced ETCs. So why have they not taken over? He argues that, in a steady-state era, knowledge will be the most valuable resource. Advanced ETCs would have a reason to leave a life-bearing planet well alone, if only because the planet will provide a non-renewable source of information. And the sacrifice of *lebensraum* need not be great. As Asimov pointed out,[62] ETCs might move beyond the need for planet-dwelling. If ETCs can travel between the stars in space arks, then they need not visit Sun-like stars; any star will do, and bright O-type stars might be best. Such space arks might therefore, on principle, avoid Sun-like stars with habitable planets. Fogg suggests the number of stars that ETCs must avoid may be small: he gives a figure of 0.6% for the fraction of stars possessing a life-bearing planet. (This figure is, of course, debatable.) Leaving a small number of systems untouched is a small price to pay for the information content their life-bearing planets will eventually possess.

In the steady-state era, then, an era in which ETCs communicate with each other and common approaches are agreed upon, the "Galactic Club" agrees not to interfere with already populated planets. In the words of Newman and Sagan,[63] a "Codex Galactica" is established. Fogg's suggestion is that the Solar System was placed under interdict when, billions of years ago, an ETC visited the Earth and discovered primitive organisms. Since then, organisms on Earth have lived in a zoo — studied for the complex patterns of information they generate.

*   *   *

To my mind, some of the premises underlying the interdict scenario are unconvincing. To take just one, I believe that the cultural homogeneity that Fogg suggests is unlikely to come to pass. I find it implausible that truly alien intelligences, if they exist, can communicate so efficiently that they reach "an enhanced level of understanding [and] mutual agreement." The problems in establishing a transgalactic communication system go way beyond mere translation difficulties. For example, the differential rotation of the Galaxy causes a star like the Sun to move relative to other stars. Fifty million years ago, Earth may have been in a region of the Galaxy in which the zoo keepers were punctilious; right now, though, we may be entering a region where the zoo keepers have evolved and decided to take some time off. If they did that, who else would know? And what could the other members of the Galactic Club do to stop it? We live in a Universe in which there is a speed limit for information flow, and it makes Galactic cultural homogeneity extremely difficult to achieve. McDonald's may have conquered the world, but it will not conquer the Galaxy.

FIGURE 19   A galaxy like this one is typically 100,000 light years or more in diameter. The interdict scenario requires a "Galactic Club" to be able to enforce its rules and traditions from one end of the galaxy to the other. In a relativistic Universe, this is extremely difficult to achieve.

So even without questioning the detailed parameters and assumptions underpinning Fogg's computer model, the conclusions are open to debate. Putting those reservations to one side, the interdict scenario suffers from some of the same criticisms as does the original zoo scenario. There seems to be no way of discovering whether we are under interdict (until, perhaps, we advance enough as a species to be elected as members of the Galactic

Club). So there are no testable predictions. The scenario also supposes that advanced ETCs, at all stages in their own evolution, can hide their activities from us. Well, maybe they can. But if the Galaxy really is teeming with ancient ETCs, as is suggested, would we not see the occasional grand astroengineered structure or overhear the occasional piece of interstellar gossip? Putting a planet under interdict is one thing; hiding all evidence of their existence from us is something else. Finally, as discussed above, even if deep communication were established in the steady-state era of the Galaxy, would a uniformity of motive regarding life-bearing planets really arise? The existence of just *one* advanced ETC that does not share the values discussed above could be enough to invalidate the scenario.

SOLUTION 7: THE PLANETARIUM HYPOTHESIS

Real are the dreams of Gods.
John Keats, *Lamia, I*

Stephen Baxter has proposed an interesting variant on the zoo scenario. He calls it the planetarium hypothesis.[64] (The speculation is far wilder than Ball's idea, but it merits the term "hypothesis" rather than "scenario" because it offers testable predictions.) Is it possible, Baxter asks, that the world we live in is a simulation — a virtual-reality "planetarium" engineered to present us with the *illusion* that the Universe is devoid of intelligent life?

The physics behind such an idea has a modern feel to it. Indeed, the planetarium hypothesis could only reasonably have been proposed in recent years — times that have seen an incredible increase in the power of computers. And yet the "things are not what they seem" concept that underlies the planetarium hypothesis is an established trope of science fiction. In Heinlein's novella *Universe*,

FIGURE 20   *In a well-designed planetarium we can lose ourselves in a realistic representation of the Universe.*

the inhabitants of a generation ship (see page 63) find a Universe beyond the confines of their vessel. In a light-hearted short story by Asimov, written two years before Soviet satellites photographed the far side of the Moon, the first astronauts to orbit the Moon find not a cratered surface but a huge canvas propped up by two-by-fours. The "trip" was a simulation that enabled psychologists to study the effects of a lunar mission on the crew. The protagonist of *The News from D-Street*, a much more somber story by Andrew Weiner, discovers that the totality of his familiar yet strangely restricted world is the product of a computer program. More recently, mainstream media have explored the concept of people interacting with various engineered realities. Several episodes of the TV show *Star Trek: The Next Generation*, for example, were set on the "holodeck" — a technology that emulated material objects with which users could interact. The movie *The Matrix* had humans forcibly immersed in a virtual reality, this time through a technology in which brains were stimulated directly by implants. The protagonist of the movie *The Truman Show* was the unwitting star of a TV show that had him living inside an engineered reality; in this case it was a "low-tech" reality, a fake town below a painted dome designed by the show's producers.[65]

Many of these stories and movies have a haunting quality, perhaps because they touch upon matters of deep philosophical concern. After all, questions about the nature of reality, and about how each of us perceives the external Universe, have kept philosophers in business for millennia. The planetarium hypothesis suggests that our accepted understanding of the external Universe might be wrong. Exactly how wrong depends on the type of planetarium the ETC has provided for us ("low-tech" like *Truman* or "high-tech" like *Matrix*) and also its scope — the position of the boundary between human consciousness and external "reality."

The planetarium hypothesis taken to extreme is similar to solipsism. The true solipsist believes that everything he experiences — people, events, objects — is part of the content of his consciousness, rather than an external reality in which we share. It is not just that his is the only mind that exists. (The sole survivor of some planet-wide catastrophe might be correct if he believed his was the only mind, and yet he would not necessarily be a solipsist.) Rather, the true solipsist in principle can attach no meaning to the idea that other minds experience thoughts and emotions. It is an egocentric view of the Universe. The most extreme planetarium, therefore, would have an ETC generate an artificial Universe directly into *my* consciousness. The Universe appears to me to be empty because an ETC, for some reason, wants to fool *me* into so thinking.

Solipsism seems to lead nowhere and is rarely defended directly. (The true solipsist when defending his philosophy presumably has to inform his opponents they do not exist, which seems a ludicrous thing to do.) Less extreme planetaria still have a solipsistic flavor but are slightly less outrageous. For example, perhaps we humans are real but some or all of the objects we see around us are simulations — like the holodeck in *Star Trek*. Or perhaps reality consists of everything on Earth plus those places in the Solar System we have visited, but the stars and galaxies are simulated — like a large-scale version of *The Truman Show* dome.

Occam's razor gives us a good reason for rejecting all these planetaria. Suppose you throw a ball and watch its parabolic path: you will conclude the ball is an autonomous object obeying Newton's law of gravity. The alternative — that some system (whether an individual consciousness or a sophisticated virtual-reality generator) contains laws that simulate the properties of the ball and its motion under gravity — is a more complex explanation of the same phenomenon. Both explanations fit the observations. But Occam's razor tells us to use the simplest explanation, which in this case is that the ball is "real." It has an autonomous existence. We can make the same argument regarding our observations of the Universe.

On the other hand, if we are willing to put Occam's razor to one side for the moment and take the planetarium hypothesis seriously, Baxter shows how we can test whether we are living in certain types of engineered reality. This is an advance on the original zoo and interdict scenarios, neither of which make hard predictions.

Baxter points out that a fundamental requirement of a planetarium is that scientific experiments should always yield consistent results. (At this point, we do not ask *why* an ETC would bother simulating a Universe for our benefit. It is enough to note that a *perfect* simulation of a system — in other words, a simulation that cannot be distinguished from the original physical system by any conceivable test — can in theory be generated.) If an experiment highlights inconsistencies in the fabric of reality, then we might be led to postulate the existence of an "outside."

Physicists can calculate the information and energy demands required to create a perfect simulation of any given size. We can therefore ask whether an ETC has the capacity to meet the energy demands for the construction of any particular planetarium. (We have to assume that the planetarium designers are subject to the same laws of physics as us. If they are not — if, for example, they can alter the value of the Boltzmann constant — then we cannot take the argument further.)

It turns out that a K1 civilization could generate a *perfect* simulation of about 10,000 km^2 of Earth's surface and to a height of about 1 km. In other words, a K1 civilization could not have generated a perfect simulation of the ancient Sumerian empire, much less our present world. Of course, a planetarium designer would not need to generate a perfect simulation in order to fool the people of Sumer; it would be unnecessary to emulate material 200 m below Earth's surface, for example, since humans of that time were unlikely to dig that deep. Various tricks and short-cuts would also be available to the planetarium programmer — but note the resulting simulation would not be *perfect*, and in principle an inconsistency might be revealed. The protagonist in Weiner's *The News From D-Street* finds himself in exactly this situation.

A K2 civilization could have generated a simulation to fool Columbus. But the voyages of Captain Cook might have uncovered inconsistencies in their planetarium design.

A K3 civilization could generate a perfect simulation of a volume with a radius of about 100 AU. This is a large distance. For comparison, Pluto, the outermost planet in the Solar System, lies at an average distance of 40 AU from the Sun; the Voyager 1 spacecraft, the most distant man-made object, is only slightly farther away than Pluto. So it is possible that humans are creatures in the simulation of some K3 civilization.

The Bekenstein Bound

Jacob Bekenstein[66] showed how quantum physics places a limit to the amount of information a physical system can code. The uncertainty relations show that the amount of information inside a system of radius R (in meters) and mass M (in kilograms) can never be greater than the mass multiplied by the radius multiplied by a constant (which has a value of about 2.5×10^{43} bits per meter per kilogram). Nature permits a surprising amount of information to be encoded before the *Bekenstein bound* is reached. For example, a hydrogen atom can encode about 1 Mb of information — most of a floppy disk. A typical human can code about 10^{39} Mb of information — far more information than can be handled by any hard disk in existence.

Natural physical systems seem to encode much less information than Nature permits. But the Bekenstein bound gives planetarium designers plenty of opportunity to engineer perfect simulations of varying size and scope. Standard thermodynamic calculations give us the energy required to construct a perfect simulation of any particular size and mass.

With our present level of technology, therefore, we are incapable of testing whether our Universe is "real" or the result of a simulation developed by a K3 civilization. But as we probe more of the Universe, and have our probes travel well past Pluto and into the outer reaches of the Solar System, we will reach a point where we can be certain that any simulation is less than perfect. A simulation could exceed 100 AU, but it would not be a *perfect* simulation; our instruments could in principle detect the inconsistencies in such a lower-quality simulation. In a few years, Voyager 1 will pass the 100-AU boundary; if it bumps into a metal wall that has been painted black — well, the game will be up for the planetarium builders!

The planetarium hypothesis defies both Occam's razor and our basic intuition about how the Universe works. It verges on paranoia to suppose that a K3 civilization would go to such effort simply to persuade us that our Universe is empty. Baxter himself advances it only as a possibility to be eliminated (and I am sure he does not believe it to be true). But at least we *can* eventually eliminate it. In the decades to come, as we explore more of the Universe and test the fabric of reality at ever-larger distance scales, we will either find an inconsistency in the simulation or be forced to accept that the Universe is "real." And if it turns out the Universe is "real" — which I am sure most readers would be prepared to wager is the case — then we will have to look elsewhere for a resolution of the Fermi paradox.

SOLUTION 8:   GOD EXISTS

Chance is perhaps God's pseudonym when he does not want to sign.
Anatole France, *Le Jardin d'Epicure*

Some have suggested SETI scientists are engaged in a theological pursuit: since ETCs are likely to be far in advance of us, we will think of them as almost omniscient, omnipotent beings. We would think of them as gods. Many SETI scientists would disagree: ETC's technology might indeed be so far advanced that it is, to use Clarke's phrase, indistinguishable from magic, but surely we know enough to consider these beings as master engineers. At worst, we would look on them as thaumaturgists. We know enough not to think of them as gods.

Others have argued that God — the creator of our Universe — exists. And that, since God is everywhere, our search for extraterrestrial intelligence would be satisfied if we found God. I am hopelessly unqualified to argue these points. However, there is a speculation from the realms of theoretical physics that might, if proved true, demonstrate the existence of many other universes that are conducive to the development of ETCs; an even more speculative suggestion is that one of those civilizations created our own Universe. They would, in a sense, be God. The work is *highly* speculative, but the theory makes a definite prediction that can be tested. The argument is as follows.

*   *   *

Physicists may be on the verge of discovering a "theory of everything": a theory that unifies gravity with the other forces and that explains the observed relationships between the various forces. (There are two points to note here. First, a "theory of everything" would answer basic physics questions. Every type of question a physicist might ask could *in principle* be answered in terms of the theory. In practice, most questions would not be explained in terms of ultimate principles, any more than present problems in protein synthesis require a knowledge of QCD for their answers. A theory of everything certainly does not have to explain love or truth or beauty. Second, physicists expressed similar sentiments about a final theory as far back as the 19th century, so we should take such announcements with a pinch of salt. But this time it really may be different.)

The present candidate for a final theory is called M-theory. The mathematics of M-theory is exceedingly difficult; indeed, much of the mathematical machinery needed to develop the theory has yet to be invented. But suppose in the next few decades M-theory is developed to a high degree of sophistication. Will it explain "everything"? Perhaps it will; that is the hope of most workers in the field. There are indications, however, that the theory will have a number of parameters — such as the masses of the fundamental particles and the relative strengths of the fundamental forces — whose values must be put into the theory "by hand." The equations of our final theory might say, for example, that the electron mass should be non-zero or that the mass associated with the cosmological constant should be non-zero — but the equations might say nothing about why those masses, in natural units, should be so tiny: 10^{-22} and 10^{-60}, respectively. It might turn out that those masses, and the various other parameters in the theory, could have taken *any* value.

If a theory of everything fails to explain why fundamental parameters have the values we observe, we would have a final theory that describes a multitude of possible universes. Each universe would have different values for the various fundamental parameters. How, then, could physicists answer a perfectly reasonable question, such as: "Why is the mass of the proton 10^{-19} in natural units when we would naively expect its mass to be about 1?" How can we proceed?

One approach is to say the parameter values were set by chance. How, though, can we explain the fact that the observed values of these parameters seem to be necessary for life? You can tinker with the parameters a little, but not much: life requires chemistry, chemistry requires stars, stars require galaxies... and all of these require the parameters to lie within a narrow range of values. Decrease the strength of the strong interaction by a factor of 4, say, and no stable nuclei can exist; we would not have stars. Change the cosmological constant by a factor of 10, say, and you end up with a universe totally unlike the one we inhabit. Lee Smolin estimates the probability of picking a set of random parameters that generate a universe favorable to life is 1-in-10^{229}. It is difficult to convey just how *fantastically* unlikely this is. For example, imagine you have a single ticket in a cosmic lottery that has roughly the same odds as the UK National Lottery: about a 13 million-to-1 winning chance. You might think it worth entering: you are not likely to win but, hey, someone has to. Now suppose the commissioners of this cosmic lottery are miserly beings. Their lottery has been drawn once a second, every second, since the start of the Universe some 13 billion years ago — so there have been roughly 10^{17} draws. But they pay out on only *one* of those draws; all other draws are void, and they keep the money. So there is only one chance in a hundred million billion that your ticket is eligible for the prize draw; and even if it is eligible, there is only a 13 million-to-1 chance that it will win. With these odds you would not bother to enter. But the chance of winning such a lottery does not even *begin* to convey the sheer improbability of a 1-in-10^{229} chance coming up. If Smolin's probability estimate is correct, then we simply cannot appeal to good luck.

A second approach is to invoke some form of anthropic principle (see page 143 for more discussion of the principle). In other words, the parameters are tuned to these unlikely values in order for rational creatures to exist. Perhaps God explicitly set the parameters to create a Universe with life; or, taking a less theological view, perhaps there are many universes, each of which has different laws and constants of physics. We then must find ourselves in a Universe where the parameters are conducive to life — after all, we can hardly find ourselves in a Universe where physics does not allow life to exist. Many scientists feel uneasy with such arguments, since anything can be explained this way; to argue like this is almost an abdication of scientific responsibility. Furthermore, a persistent criticism of the anthropic approach is that, with a couple of debatable exceptions, it fails to make predictions that can be tested by observation.

A third approach, promoted by Smolin, is to apply Darwin's evolutionary ideas to cosmology.[67] Equations may not be able to explain why physical parameters have fine-tuned values like 10^{-60}, *but evolutionary processes can*. Smolin suggests that the physical constants — and perhaps even the laws of physics — have evolved to their present form through a process that is similar to mutation and natural selection.

How can this be? Smolin's key assumption is that the formation of a black hole in one universe gives birth to another, different expanding universe. He further assumes that the fundamental parameters of the child universe are slightly different from those of the parent universe. (This process is thus rather like mutation in biology: the child has a similar genotype to the parent, but there can be a slight variation.) In this picture, the Universe we live in was generated through the formation of a black hole in a parent universe with similar physical constants to our own. A universe with parameters that permit the formation of black holes has offspring that will in turn produce black holes. A universe with parameters that lead to little or no black-hole formation will produce little or no offspring. Very quickly, no matter how fine-tuned the parameters need to be, universes with parameters that lead to black-hole formation will come to dominate: pick a universe at random and the chances are overwhelming that you pick a universe in which many black holes form.

Now, so far as we know, the most efficient way for a universe to produce black holes is through the collapse of stars. For example, our own Universe will create as many as 10^{18} black holes — and thus, in Smolin's picture, child universes — through stellar collapse. So, no matter how "improbable" the values of the fundamental physical parameters that allow stars to form, we expect cosmic evolution to generate a preponderance of universes in which there are innumerable stars. And a universe with physical parameters that gives rise to stars is a universe that inevitably has heavy nuclei, and chemistry, and long enough timescales for complex phenomena to emerge. In other words, it is a universe that may have life. The fine-tuning of the constants is for the benefit of black-hole production rather than the production of life. In Smolin's picture, life is simply an incidental consequence of a universe that has sufficient complexity to allow the formation of black holes.

FIGURE 21   An artist's impression of the supermassive black hole in MCG-6-30-15, a distant galaxy. Astronomers believe the cores of most galaxies contain supermassive black holes — perhaps each of these black holes creates a universe with physical parameters like our own? If so, our Universe may have given rise to billions of similar universes. Even more common than supermassive black holes are those formed in stellar collapse. If these objects create new universes, then our own Universe may have 10^{18} offspring!

This may sound like speculation, and it is. Indeed, the idea is almost entirely speculative. There is no evidence (and perhaps there never can be) that the formation of a black hole creates a different expanding universe. Even if a new universe does form, we cannot answer many of the questions we would like to ask. (Exactly how do the physical parameters change at the birth of each child universe? Does a single black hole always give rise to a single universe? Does the mass of the black hole play any role? What happens if several black holes merge? And so on, and so on.) Until we have a quantum theory of gravity, we cannot begin to attack such questions. Nevertheless, Smolin's idea has a certain attraction: it links key scientific ideas — evolution, relativity and quantum theory — to explain the long-standing puzzle of the values of the fundamental parameters of physics. Moreover, it makes a specific forecast, a prediction against which the theory can be tested. The prediction is that, since we live in a Universe that creates many black holes and can therefore assume that the fundamental parameters are close to optimum for black-hole formation, a change in any of the fundamental parameters would lead to a Universe with fewer black holes.[68]

In a few cases, physicists have been able to calculate what would happen if a fundamental parameter differed from its observed value. In each case, it would indeed lead to a reduction in the number of black holes formed by stellar collapse. At present, though, we do not understand enough about astrophysics to calculate the effects of varying all the parameters. Smolin's idea is neither ruled in nor ruled out; it remains an intriguing speculation.

*   *   *

Edward Harrison takes the speculation one step further.[69] He too highlights the long-standing puzzle of why the physical constants seem to be *just* right for the development and maintenance of organic life. Smolin's theory goes part of the way to explaining the puzzle, but Harrison argues that the link between black-hole formation and the conditions necessary for life is too tenuous. Suppose, though, some time in the future, Smolin's idea transmutes into established cosmological theory. Then, Harrison suggests, we might come to believe we should make as many black holes as possible, for in doing so we would increase the probability that other universes might contain intelligent life. If in the far future *we* might create child universes, perhaps our *own* Universe was created by intelligent life. Perhaps God did not labor for six days; maybe it was an ETC, in a universe with fundamental physical parameters much like our own, that labored to create a black hole — a black hole that led to the formation of our Universe.

I am not sure if Harrison's suggestion could ever resolve the Fermi paradox. Could the ETC squeeze some sort of message through the bounce that creates another universe? If not, how could we ever know whether our Universe was artificially produced in a laboratory inside some other universe? The notion that they could squeeze through a message is, however, intriguing. Even if it happened that *our* Universe was devoid of other intelligent life, we would at least know we were not alone ... sort of, anyway.[70]

4

They Exist
But Have Not Yet
Communicated

The position many scientists take on the question of extraterrestrial life is this: the Galaxy contains tens of thousands of life-bearing planets, and on some of those planets ETCs exist that are technologically far in advance of our own. This conclusion seems to follow from the Principle of Mediocrity — the notion that Earth is a typical planet orbiting a common type of star in an ordinary part of the Galaxy. The principle has served science well since the time of Copernicus. Scientists who take this position, however, have to answer Fermi's question. If ETCs exist, then why are they not here? At the very least, why have we not heard from them?

There are a variety of answers, ranging from the technological (interstellar travel is difficult to achieve, for example) to the sociological (for example, all societies sufficiently advanced to develop interstellar travel inevitably destroy themselves). One weakness of many of these answers, particularly sociological answers, is that to explain the Fermi paradox they must apply to *every* ETC. I leave the reader to decide whether such answers can resolve the paradox, either singly or in combination.

Solution 9: The Stars Are Far Away

...between stars, what distances.
Rainer Maria Rilke, *Sonnets to Orpheus*, Part 2, XX

Perhaps the most straightforward solution to the Fermi paradox is that the distances between stars are too great to permit interstellar travel. Perhaps, no matter how technologically advanced a species becomes, it cannot overcome the barrier of interstellar distance. (This might explain why ETCs have not *visited* us, but not necessarily why we have not *heard* from them. But let us put this criticism to the side for the next few sections.)

That the stars are far away does not in itself make interstellar travel unattainable. It is not impossible to build a vessel that can leave a planetary system and then travel through interstellar space. Take our Solar System as an example: its escape velocity, starting at Earth's distance from the Sun, is only 42 km/s. In other words, if we launch a vessel traveling at 42 km/s relative to the Sun, then it can escape the grip of the Sun's gravitational influence. It can become a starship. No problem: NASA has already built several such vessels! (With our present technology we have to cheat a little and use the gravity assist offered by the planets. The so-called "slingshot effect" is quite sufficient to boost a slow-moving craft to escape velocity.)

Voyager 1, launched on 5 September 1977, toured the outer planets and then headed out into space. On 17 February 1998 it became the most distant man-made object, and it is now farther from the Sun than is Pluto. Unless alien probes pick it up, as happened to the fictional Voyager 6 in *Star Trek: The Motion Picture*, it will eventually make its closest approach to a star — it will drift within 1.6 light years of an unprepossessing M4 star called AC +79 3888. The trouble is, Voyager will take tens of thousands of years to reach its closest encounter with the star. And *that* is the difficulty with interstellar travel: unless you travel fast, the transit times are long.[71]

The best way to rate a starship's speed is in terms of c, the speed of light, since c is a universal speed limit.[72] The speed of light in a vacuum is 299,792.458 km/s. So Voyager 1, which as I write is traveling at 17.26 km/s away from the Sun, travels at a mere $0.000058c$. Now, the stars are so widely separated that a favored method of presenting interstellar distances is to use the *light year*: the distance light travels in one year. For example, the nearest star to our Sun is Proxima Centauri, which is 4.22 light years distant.[73] So the fastest possible "craft" — photons of light — would take more than 4 years to reach the nearest star; Voyager 1, were it traveling in that direction, would take almost 73,000 years to complete the same journey. The huge travel time involved when traveling at sub-light speed leads many commentators to conclude that interstellar travel, while perhaps not theoretically impossible, is impracticable.

But perhaps exploration of the Galaxy, even at Voyager speeds, is possible. As we have seen (page 45), the notion of directed panspermia supposes that the Galaxy could be seeded with life using slow-moving probes. And as long ago as 1929, John Bernal proposed the idea of the "generation ship" or "space ark": a slow-moving self-contained craft that would effectively constitute the whole world for its passengers. After setting off from the home planet, many generations of passengers would live and die before the craft arrived at its destination.[74] Bernal's idea was wonderfully dramatized in Heinlein's story *Universe*.[75] Another possibility would be to put the passengers into suspended animation, as in the film *Alien*, and revive them upon arrival. It has even been suggested that frozen embryos could be transported on slow-moving craft, and then grown in artificial wombs at journey's end.

Clearly, though, if we wish to reach the stars in a reasonable time, we need to build craft that can travel at a substantial fraction of the speed of light. Even then, the travel times involved may be long on an individual human scale. For example, ignoring the acceleration and deceleration times at either end of a journey, a craft traveling at the enormous speed of $0.1c$ would take 105 years to reach Epsilon Eridani, which is one of the nearest Sun-like stars. Few crew members seeing their new star for the first time would remember the star their ship left behind. (When talking about

FIGURE 22   *The 110-m-tall Apollo 11 spacecraft was launched from Pad A, Launch Complex 39, Kennedy Space Center, at 9:32 A.M., 16 July 1969. On board were astronauts Armstrong, Aldrin and Collins. This vehicle, the first to land men on another world, would be impractical for interstellar travel.*

travel times, we tend to assume that people will choose *not* to spend so many years of their life away from home. But we base this assumption in terms of the present human lifespan. After gaining their degrees, several of my more adventurous contemporaries chose to spend a year — which is roughly 2% of their adult life — simply traveling around the world. If human lifespans were increased by a factor of ten, say, then perhaps an adventurous soul would be quite willing to spend a mere decade of his life traveling to the stars. Perhaps even a century-long journey would not be uncommon. Who knows? As always, it is difficult to argue about future activities based on present technology.)

The journey time mentioned above — 105 years to reach Epsilon Eridani, at $0.1c$ — is the time that Earthbound observers would measure. People on the ship would measure a slightly smaller interval due to the special relativistic effect of time dilation.[76] We are justified in ignoring time dilation effects for on-board observers traveling at $0.1c$, since the effect is only about 0.5%. The closer the speed is to c, however, the more noticeable the effect. A craft traveling to Epsilon Eridani at $0.999c$ would take 10.5 years to complete the journey as measured by Earthbound observers, but to a crew member the journey would take only 171 days! If it were possible to travel at speeds infinitesimally smaller than c, then *for the traveler* the journey would take a mere fraction of a second. A trip to the farthest galaxies would be possible within a human lifetime — though to Earthbound observers the trip would take so long that Earth itself would be consumed in the Sun's death throes.[77]

What is the likelihood that an intelligent species could develop techniques for interstellar travel at reasonable speeds? (By "reasonable" I mean any speed that enables a mission to reach nearby stars on a timescale of hundreds rather than tens of thousands of years. Highly relativistic speeds would be nice, of course, since they would put the stars within reach of individuals living a human lifespan. But a craft leaving the Solar System traveling at $0.01c$ will reach the nearest star in about 430 years, which puts the stars within range of generation ships.) To answer this, we need to consider the various space-travel technologies that have been suggested. I give only a brief overview here; the notes in Chapter 7 point to further resources.

Although I concentrate here on propulsion methods, it is worth bearing in mind that there are other factors to consider. For example, a starship traveling at high speeds would suffer a ferocious bombardment — tiny dust particles from the interstellar medium would deposit large amounts of energy into the starship structure. Protecting the structure against such erosion, and protecting the crew from the more insidious problem of cosmic-ray bombardment, would require sophisticated shielding. There is also a navigation problem: the stars move with different velocities in three dimensions, making it difficult for a slow-speed mission to rendezvous with a particular star.[78] Nevertheless, these problems are moot if no systems exist that can propel a ship to the stars. If interstellar travel is impossible, then maybe we have a solution to the Fermi paradox.

Rockets

Most people's initial idea for a starship propulsion mechanism is the self-contained rocket. NASA's familiar chemical rockets obtain all their energy and expellant mass from on-board reserves. Consider the Apollo missions, for example. The multi-stage Saturn V rockets burned liquid propellants: a mixture of kerosene with liquid oxygen for the first stage, and liquid hydrogen with liquid oxygen for the second stage. The exhaust from these chemical reactions was sufficient for reaching the Moon, but this approach is simply not feasible for interstellar travel: the nearest star is more than 100 million times more distant than the Moon. The kerosene tanks would be enormous!

Nevertheless, it may be possible to employ variations on this theme. For decades, scientists have considered alternatives to chemical rockets. An ion rocket, for example, would expel charged atoms to generate thrust; a nuclear fusion rocket would generate high-speed particle exhaust by means of controlled thermonuclear reactions. Perhaps the boldest possibility is the antimatter rocket, first suggested in 1953 by Eugen Sänger. When a particle of matter comes into contact with its antiparticle, both particle and antiparticle mutually annihilate and produce energy. Choose the initial particles correctly and it might be possible to channel the annihilation products into a directed exhaust. Although further analysis showed that Sänger's initial design could not succeed, advances in antimatter physics made in recent decades have stimulated proposals that *may* one day lead to an antimatter rocket.[79]

Fusion Ramjets

The whole concept of using a self-contained rocket — which has to carry the energy source *and* the payload — may be impractical for interstellar travel. Are there propulsion systems that do not require the ship to carry its own fuel? In 1960, Robert Bussard suggested that a *fusion ramjet* might power its way to the stars.[80]

The space between stars is not empty. There exists an interstellar medium, comprised chiefly of hydrogen. A ramjet would use an EM field to scoop up this hydrogen and funnel it to an on-board fusion reactor, which in turn would "burn" the hydrogen in thermonuclear reactions to produce thrust. As with Sänger's antimatter rocket design, Bussard's fusion ramjet proposal suffers from a host of practical difficulties. It is unlikely that Bussard's initial idea could be made to work. Nevertheless, several studies have proposed methods to improve the design. Perhaps one of these designs could eventually form the basis of a working starship. Enthusiasts remain enticed by the possibility of the ramjet, because in theory it could attain speeds close to *c* after just a few months.

Chapter 4

Laser Sails

At about the same time that Bussard proposed the fusion ramjet, Robert Forward proposed the *laser sail* as a means of reaching the nearest stars.[81] Imagine a vast "sail" attached to a spaceship; and imagine a giant solar-powered laser aiming a narrow beam of radiation toward the ship. Photons from the beam would cause a tiny pressure on the sail, and the ship would be gently pushed toward the stars. A laser sail could accelerate to extremely high velocities; hitting the brakes would be more difficult, although deceleration mechanisms have been proposed. Forward's idea has been refined over the past four decades, and enthusiasts have designed schemes to use laser sails for both a one-way colonization mission and a round-trip to the stars.[82]

FIGURE 23   *This beautiful painting shows a solar-powered space-based laser focusing a beam on the huge lightweight sails of a spacecraft.*

Gravity Assists

In 1958, Stanislaw Ulam considered the possibility of accelerating a ship to high velocity using its gravitational interaction with a system of two much larger astronomical bodies in orbit around each other. (It is a trick similar to the gravity-assist trajectories that gave Voyager 1 sufficient velocity to leave the Solar System.) A few years later, Freeman Dyson considered more realistic (though still, of course, speculative) scenarios. Using Dyson's approach, an advanced technological civilization might employ two orbiting neutron stars to accelerate spaceships to near light speed.[83]

Fancy Physics

The technologies mentioned above are based on established physics. The construction of starships using these ideas are, of course, way beyond our present capabilities; indeed, engineering considerations may make it impossible *in practice* to construct starships. But there seems to be nothing wrong with these ideas *in theory*. They break no physical laws.

For many years, people have wondered whether it is possible to travel *really* fast. If we could travel at speeds greater than c, then the stars would no longer be grindingly distant. Faster-than-light (FTL) travel would bring the ends of the Galaxy within reach. Nearly all ideas for FTL travel can immediately be discounted, since they clearly violate established physical principles. A few suggestions, however, have not yet been ruled out.

Tachyons. The special theory of relativity does not absolutely forbid superluminal travel. Rather, it states that massive particles cannot be accelerated to light speed, while massless particles (like photons) always travel at the speed of light. Particles with *imaginary* mass must always travel *faster* than the speed of light. Such imaginary-mass particles are called tachyons.

There is nothing particularly unusual about imaginary quantities: we represent several physical quantities by imaginary numbers. But it is difficult to understand what an imaginary *mass* represents. We have no problem understanding the idea of a positive mass; nor is there any difficulty with the idea of a zero mass; we can even ascribe meaning to negative mass (and note that, if negative mass existed, we might be able to use it in a propulsion device).[84] But *imaginary* mass? Whatever it might mean, physicists have searched for signs of it. So far, the tachyon remains hypothetical. There is no evidence such particles exist, and our theories work fine without them. Even if we found tachyons, how could we harness them for FTL travel? We are clueless, here, and it seems reasonable to strike tachyon drives from the list of propulsion possibilities.

Wormholes and warp drives. Most of us are familiar with the Newtonian picture of gravity. We are taught in school that massive objects attract one another by exerting a mysterious influence through empty space. Einstein's general theory of relativity presents a very different picture of gravity. In this view, space — or rather, spacetime — plays an active part in the gravitational interaction. In the words of John Wheeler: mass tells spacetime how to curve, and curved spacetime tells mass how to move.

We can think of special relativity as a particular case of general relativity. It applies locally to any region of spacetime small enough that its curvature may be neglected. The interesting point to consider here is that general relativity permits FTL travel — so long as the *local* restrictions of special relativity are obeyed. The speed of light is a local speed limit, but general relativity permits ways to circumvent this limit. Although this may seem peculiar, there are well-established examples of FTL phenomena in general relativity. For example, standard cosmological models suggest that, due to the expansion of the Universe, distant regions of space recede from us at FTL speeds. Only if the expansion slows will those regions appear over the light speed horizon and be visible to us.

So far, general relativity has passed every experimental test. It correctly predicts the bending of light rays near the limb of the Sun, the orbits of binary pulsars, and the arrival of signals in GPS systems. However, most tests of the theory occur in situations where spacetime curvature is small. Sometimes, the distribution of matter can cause a large curvature of spacetime. At the singularity of a black hole, for example, the density of matter is infinite; the very fabric of spacetime is punctured.

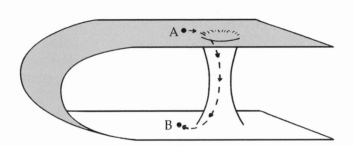

FIGURE 24 *If space folds over on itself, then a wormhole linking A to B might allow travelers to move between these points without having to traverse the "normal" spacetime between the points.*

It is difficult to interpret the results of general relativity in the extreme situations that occur near the singularity of a black hole. Perhaps the theory cannot be applied in such situations; we may require a quantum theory of gravity to describe what happens there. But in an attempt to understand these extreme regions of spacetime, physicists have pushed the theory. One speculation is that the formation of a black hole can lead to the formation of a *wormhole* — a "bridge" that links two separate black holes. The two holes may link two quite separate points of spacetime, or two different regions of the Universe. Enter one black hole and you might emerge from the other hole moments later, thousands of light years from your starting point. As you traveled through the bridge you would have observed the local speed limit and moved slower than c; yet your effective speed could be millions of times greater than c. Sagan used this idea in his SF novel *Contact*.[85]

Although based on solid work, the wormhole remains a hypothetical creature in the theoretical physicist's bestiary. Wormholes may not exist. Even if they *do* exist, we may be unable to travel through them: calculations suggest that they are likely to be small and wildly unstable. Nevertheless, there remains a tantalizing possibility that an ETC in possession of "exotic" matter (matter with a negative mass–energy) could take a microscopic wormhole, stabilize it, inflate it to a large size — and then use it to traverse huge distances. Recently, the Russian physicist Sergei Krasnikov has shown that a certain class of wormhole might be constructed using "normal" (positive mass–energy) matter. Perhaps a K3 civilization could use such Krasnikov wormholes for interstellar travel.

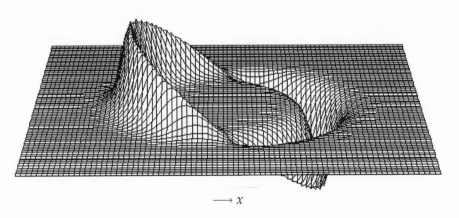

$$\longrightarrow x$$

FIGURE 25   *The figure shows the curvature of space in the region of Alcubierre's warp. Space expands at the rear of the warp and contracts at the front; the flat region is pushed forward.*

There is another way in which general relativity might permit superluminal travel (and in the style to which *Star Trek* has accustomed us). Imagine a spaceship — one as large and luxurious as the QE2 — inside a flat region of spacetime. Everything on board the ship would behave as it does in the flat region of spacetime we are accustomed to here on Earth. Now imagine that, at the rear of the volume, space expands (in the same way that the Universe itself expands). And at the front of the volume, space contracts (as would happen if the Universe were to collapse into a Big Crunch). The result of this particular warp in space is that the flat-space volume, containing the spaceship, would move forward — propelled by the expansion of space at the rear and the contraction of space at the front. The ship effectively surfs a spacetime wave.[86]

The warp can travel at arbitrarily large speeds, perhaps many times faster than c, and it carries the ship with it. With respect to the local volume of flat space, however, the ship is at rest. There is no relativistic mass increase, and no time dilation. For the crew, everything is as normal. As they speed toward the stars at a speed of $100c$, the passengers are free to enjoy the hospitality of the Spaceship *QE2*.

The properties of this peculiar solution to Einstein's equations were first analyzed by Miguel Alcubierre while he was at Cardiff University. I have a soft spot for the Alcubierre warp drive, since I was wasting time in the office opposite Miguel while he was working on his idea. Nevertheless, the Alcubierre drive, at least as first proposed, is unlikely to work. First, we have no practical idea of how to produce the required curvature of space. Second, the energy density within the warped region is *very* large, and negative. (Some theorists would argue this second problem kills the whole idea of a working Alcubierre drive. However, quantum theory provides circumstances in which a negative energy density can occur. If we ever advance to the stage where we can produce large quantities of exotic matter, then *perhaps* we could make an Alcubierre drive. Even this seems unlikely, though. A warp large enough to carry the Spaceship *QE2* would require a total negative energy that is ten times larger than the positive energy of the entire visible Universe!) The Belgian physicist Chris Van Den Broeck may have found a way around some of the problems of the Alcubierre drive. The construction of a microscopically small warp bubble would require just small amounts of exotic matter; combine this with some topological gymnastics, which are allowable in general relativity, and you can end up with an interior volume of the warp bubble that is large enough to hold a spaceship. It would be rather like the Tardis in *Dr. Who*: microscopically small on the outside, but roomy enough for passengers on the inside. We may find, when we have a full quantum theory of gravity, that the Van Den Broeck drive is ruled out; in any case it is worth emphasizing that the drive is speculative and possesses unrealistic features (unreasonably large energy densities are required, for example).[87]

Perhaps wormhole and warp-drive transportation will never be practical. But they have not yet been shown to be impossible. Maybe one day.

Zero-point energy. The quantum uncertainty principle tells us we cannot know simultaneously both the position and the momentum of a particle.

Therefore even at absolute zero a particle must jitter, since if it were at a perfect standstill we would know both its position and momentum. Energy and time also obey the uncertainty principle; similarly, then, a volume of empty space must contain energy (since to establish that the energy was zero we would have to take measurements for eternity). The Casimir effect[88] — a small attractive force that acts between two uncharged parallel conducting plates brought into close proximity — is the clearest example of the existence of *zero-point energy* (ZPE). The effect can only be explained in terms of quantum fluctuations of the electromagnetic field.

Some writers suggest there is an infinite supply of energy in the vacuum and that some day we will tap into this ZPE. Perhaps we can use ZPE for a propulsion system. Recently, NASA even sponsored a meeting on innovative propulsion systems in which ZPE was identified as a potential breakthrough technology. If it works, then we have limitless cheap energy. Personally, I remain highly skeptical of the idea; we never get something for nothing. But it is yet another suggestion of how an advanced ETC might use the possibilities inherent in the laws of physics to develop technologies that seem almost magical to beings at our level of development.

*   *   *

I have only touched on the various proposals for interstellar propulsion systems. At present, we could not build *one* of the devices mentioned above and use it to reach the stars. With our present level of technology, we would find it almost impossible to send people safely to Saturn and back, let alone Sirius. There is a host of problems — economic, political, scientific and technical — that we (and presumably an ETC) would have to overcome in order to travel to the stars. What is remarkable, though, is the number of methods that reputable scientists have proposed for starflight. The methods range from the slow to the essentially instantaneous; from the tried-and-tested to the exotic. Although the human race cannot build a starship in 2002, what about in 2102? What about in 3002? Remember that 1000 years corresponds to only 2.5 seconds of the Universal Year. Other civilizations might be millions, even billions, of years older than our own. Is it likely that *none* of them have the requisite technological skill (or, if relativistic travel is impossible, simply patience) for space travel?

The stars are indeed distant. This fact alone may explain why we have not been visited (though it does not necessarily explain the "great silence" — the absence of signals from ETCs — nor why we see no other evidence of advanced civilizations). However, for those who are optimistic about the reach of science and technology, the distance barrier can be overcome. For those people, the size of the Galaxy alone does not explain the Fermi paradox.

Chapter 4

SOLUTION 10:   THEY HAVE NOT HAD TIME
TO REACH US

Had we but world enough, and time.
Andrew Marvell, *To His Coy Mistress*

A common reaction when people first hear of the Fermi paradox is: "Oh, they have not had time to reach us." Hart, in his influential paper on the absence of ETCs, called this the temporal explanation of the paradox.

As we saw on page 4, Hart argued that this explanation is not tenable. To recap, he reasoned that if an ETC sends colonization ships to nearby stars at a speed $0.1c$, and if the colonies in turn send out their own colonization ships, then that ETC would quickly colonize the Galaxy. If the ships did not pause between trips, then a colonization wavefront would sweep through the Galaxy at a speed of $0.1c$. If the time *between* voyages was about the same as the voyage time itself (travelers have to rest, after all), then the colonization wavefront would move at $0.05c$, so it could travel from one end of the Galaxy to the other in 0.6 to 1.2 million years. For ease of use, we can say that the Galactic colonization time is 1 million years.[89]

One million years is a long time on an individual level; it is a long time even at the level of an entire mammalian species. But it extremely short compared to the total time available for colonization. Consider the various time scales involved in terms of the Universal Year. The Galactic colonization time corresponds to just 41 minutes 40 seconds — less than one half of a soccer match. On this timescale, civilizations may have been popping into existence since the late spring months, and there seems to be no compelling reason why the first ETC could not have arisen by about May Day. So although the first species with the inclination and ability to engage in interstellar travel might have arisen at any time in the 8 months between May and December, according to Hart the temporal explanation asks us to accept that this species started traveling no earlier than 11:18 P.M. on 31 December. It would be a remarkable coincidence if mankind emerged so soon after the emergence of the first star-faring civilization.

Hart's argument is compelling, but one can dispute a number of his assumptions. An obvious problem is the speed of the colonization wavefront, which Hart assumes to be a large fraction of the speed of individual spacecraft. As Sagan pointed out: "Rome was not built in a day — although one can cross it on foot in a few hours." In other words, for the city of Rome, the speed of the "colonization wavefront" was an infinitesimal fraction of the speed of the craft used to "colonize" it. More explicitly, throughout all of human history there has never been a colonization wavefront that moved anything like as fast as the speed of individual craft. Why should it be any different for a civilization busy colonizing the Galaxy?

Hart calculated his Galactic colonization time simply by dividing the diameter of the Galaxy by an assumed travel speed. Several authors have developed more sophisticated computer models of Galactic colonization and thereby arrived at more plausible colonization times. Eric Jones analyzed a model in which colonization was driven by population growth.[90] He assumed a population growth rate of 0.03 per year and an emigration rate of 0.0003 per year (which was the emigration rate during the European colonization of North America in the 18th century). His model showed that, under these assumptions, a single space-faring ETC could colonize the Galaxy in 5 million years. In subsequent analyses he offered a preferred colonization time of 60 million years (though this time can be made larger with different assumptions for the rates of emigration and population growth). Of course, 60 million years is much longer than Hart's colonization time; but it is still too short to permit a temporal explanation of the Fermi paradox. On a human scale, a process that takes 60 million years is not even glacially slow; but on a cosmic scale the colonization wave moves like a flash flood through the Galaxy.

However, Jones himself made assumptions that can be disputed. For example, Newman and Sagan argued that Galactic colonization cannot be driven by the demands of population growth.[91] Look at mankind. In the last century, the world population more than tripled in size. If the population were to continue to grow at that rate, and if we wished to maintain Earth's present population density, then in a few hundred years a colonization wavefront would be moving at light speed. Once we reached that point, the population growth rate would *have* to decline! This is an extreme example, but it demonstrates that ETCs will not establish colonies as a means of avoiding overcrowding on the home planet. In the long run, they cannot outrun the problems caused by an exponentially increasing population — they simply cannot travel fast enough. A civilization has to curb its population growth regardless of whether it develops space travel. Newman and Sagan therefore modeled Galactic colonization as a *diffusion process*,[92] and applied the well known mathematics of diffusion to a particular colonization model. Their results seemed to show that if ETCs practice zero population growth, then the *nearest* civilization would reach Earth only if it had a lifetime of 13 billion years. This *is* long enough to provide a temporal explanation of why extraterrestrials are not here (though it does not necessarily explain why we do not hear from them).

The Newman–Sagan model was subject to criticism. In their model, it turns out that the Galactic colonization time is rather insensitive to the speed of interstellar travel. What matters is the time taken to establish a planetary colony, which in turn depends upon the population growth rate. Newman and Sagan assumed *very* low population growth rates — rates that many people find too conservative. Even if one accepts their rates for population growth, there is a problem with their conclusion. The differential rotation of the Galaxy turns the expansion zone into a spiral, rather like path of a drop of thick cream when you slowly stir it into a cup of coffee. Take this factor into account and the Galactic colonization time shortens dramatically. A final criticism: even if advanced ETCs are not driven to expansion by population pressure, would they not explore the Galaxy out of curiosity?

Yet other models have been analyzed.[93] For example, a recent calculation by Ian Crawford suggests that the Galaxy can be colonized in as little as 3.75 million years. The biggest uncertainty in this figure is not the speed of interstellar spaceships, but the time it takes for colonies to establish themselves and then send out their own spaceships. And Fogg, in developing his interdict scenario, analyzed the results of a model in which ETCs arise at the rate of 1 every 1000 years, and 1 in 100 of these ETCs attempts to colonize the Galaxy. His model provided the time to "fill" the Galaxy for different speeds of colonization wavefront. Even under the most pessimistic assumptions, he found that ETCs filled the Galaxy in 500 million years, which is short compared to the age of the Galaxy and makes it difficult to support a temporal explanation of the paradox.

SOLUTION 11: A PERCOLATION THEORY APPROACH

All things flow; nothing abides.
Heraclitus

The colonization models described previously address the Fermi paradox in terms of the time it might take one or more ETCs to spread throughout the Galaxy. The most recent colonization model, proposed by Geoffrey Landis, presents a more interesting solution to the paradox. Landis bases his model on three key assumptions.[94]

First, he assumes that interstellar travel is possible but difficult. No dilithium crystals; no warp engines; no USS *Enterprise* boldly going; just a long, slow haul to the closest stars. As we have seen, this is a reasonable assumption: to the best of our certain knowledge, the laws of physics do not forbid interstellar travel, but they make it time-consuming and costly. Landis thus argues that there is a maximum distance over which an ETC can establish a colony directly. Mankind, for example, may one day establish a colony directly around Tau Ceti (just under 12 light years distant from Earth) but find it impossible to directly colonize any of the stars in the Hyades cluster (150 light years distant from Earth). Any given ETC will find there is only a small number of stars both suitable for colonization and within the maximum travel distance from its home planet. Therefore any given ETC will establish only a small number of direct colonies. More distant outposts can be settled only as secondary colonies.

Second, since interstellar travel is so difficult, Landis assumes a parent civilization will possess only weak (and possibly non-existent) control of its colonies. If the timescale over which a colony develops its own colonization capability is long, then every colony will possess its own culture — a culture independent of the colonizing civilization.

Third, he assumes a civilization will be unable to establish a colony on an already colonized world. (This is tantamount to saying that invasion is unlikely over interstellar distances, which seems reasonable. If interstellar travel is both difficult and costly, then invasion must be even more difficult and more costly. There goes the plot of several Hollywood blockbusters.)

Finally, he proposes a rule. A culture either has a drive to colonization or it does not. An ETC possessing such a drive will definitely establish colonies around all suitable stars within reach. An ETC having no uncolonized stars within reach will, of necessity, develop a culture lacking the colonization drive. Therefore any given colony will have some probability p of developing into a colonizing civilization, and a probability $1 - p$ of developing into a non-colonizing civilization.[95]

These three assumptions, plus the rule, generate a *percolation* problem. The key task in a percolation problem is to calculate, for a specific system, the probability that there is a continuous path from one end of the system to the other. The word "percolation" comes from the Latin phrase meaning "to flow through," and those who developed percolation theory perhaps had in mind coffee percolation when they named it: to make a drink, water must find a path through the ground coffee and into the pot. Coffee-making is a particular example of the general problem of the diffusion of liquid through a porous solid; but percolation models have also been used to study phenomena as diverse as the propagation of forest fires, the spread of contagious disease in a population, the formation of stars in spiral galaxies, and the behavior of quarks in nuclear matter.[96]

In essence, percolation is merely a way of filling a large array of empty spaces with objects. (Strictly, percolation theory is valid only for arrays that are infinitely large, so the systems of interest must be large for percolation theory to apply.) The array need not be rectangular, nor need it be two-dimensional: some phenomena are best modeled with a one-dimensional array, others with a three-dimensional array, and still others with higher-dimensional arrays. To fix ideas, though, it is easiest to imagine a large two-dimensional array of N cells, rather like an extended chessboard.

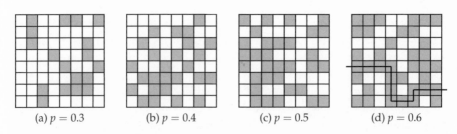

(a) $p = 0.3$ (b) $p = 0.4$ (c) $p = 0.5$ (d) $p = 0.6$

FIGURE 26 *The cells in each of these four arrays have been shaded (occupied) at random. In (a), each cell has a 30% chance of being occupied. In (d), each cell has a 60% chance of being occupied. Even in (a) there are "clusters" — cases where two or more nearest-neighbor cells are occupied. (The nearest neighbor of a cell is one that is directly above, below, left or right of the cell.) In (d) there is a "spanning cluster": a path through nearest neighbors from one end of the array to the other.*

Percolation Theory

Suppose each cell of an array has a probability p of being populated. Each cell is independent of the others — just because a particular cell happens to be populated does not mean that its neighboring cells are more or less likely to be populated. Clearly, $p \times N$ of the cells will be populated and $(1 - p) \times N$ will be empty. If the probability p is large, then the array will contain lots of filled cells; if p is small, then the array will be sparsely populated. Figure 26 shows four computer-generated 8×8 arrays. In (a) the probability of occupancy for a cell is 30%; in (b) it is 40%; in (c) it is 50% and in (d) it is 60%. (Physicists deal with much larger simulations than this, of course, but an 8×8 grid is fine for the purposes of illustration.) Two occupied cells that are next to each other are called *neighbors,* and groups of neighbors are called *clusters.* For the two-dimensional array shown in the illustration, each cell, except those on the edges, can have four neighbors: the cells directly above and below, and to the left and right. Percolation theory deals mainly with how these neighbors and clusters interact with each other, and how their density affects the particular phenomenon being studied. A cluster that spans the length or width (or both) of an array is particularly important in percolation theory. It is called the *spanning cluster,* or *percolation cluster.* For an infinite lattice, a spanning cluster occurs only when the probability p is above a critical value p_c.[97]

What has this to do with the Fermi paradox? Well, if Landis is right, we can use the well-honed techniques of percolation theory to simulate the flow of ETCs through the Galaxy. Although percolation problems are difficult to study analytically, they can be easily simulated on computer.

Readers with some programming expertise can set up the Landis model and study for themselves the distribution of ETCs under different model parameters. Figure 27 shows a typical result.

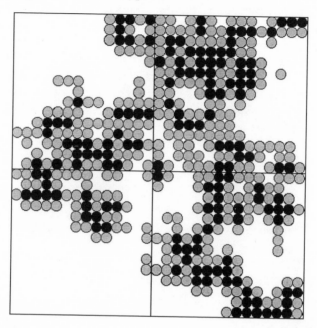

FIGURE 27   *A slice from a typical percolation simulation on a simple cubic lattice in three dimensions. For this array* p_c = 0.311, *while the simulation is for* p = 0.333. *The black circles denote "colonizing" sites, and the gray circles denote "non-colonizing" sites. The absence of circles denotes sites that have not been visited. Note the irregular shape of the boundary and the large voids. Does Earth perhaps lie in one of the voids?*

As in any percolation problem, the final lattice depends upon the relative values of p and p_c. In the Landis model, if $p < p_c$, then colonization will always end after a finite number of colonies. Growth will occur in clusters, and the boundary of each cluster will consist of non-colonizing civilizations. If $p = p_c$, then the clusters will show a fractal structure, with both empty and filled volumes of space existing at all scales. If $p > p_c$, then clusters of colonization will grow indefinitely, but small voids will exist — volumes of space that are bounded by non-colonizing civilizations. We produce a Swiss-cheese model of colonization: civilizations span the Galaxy, but there are holes.

The percolation approach thus suggests that colonizing extraterrestrials have not reached Earth for one of three reasons. First, $p < p_c$, and any colonization that has taken place stopped before it reached us. Second, $p = p_c$, and Earth happens to be in one of the large uncolonized volumes of space that inevitably occur. Third, $p > p_c$, and Earth is in one of the many small unoccupied voids. Which explanation is most probable? To answer this we need to know the value of the colonizing probability p and also the typical number of stars available for colonization. Of course, we have absolutely no idea of what a reasonable value for p might be; Landis takes $p = \frac{1}{3}$, which is as good as any other estimate. As for colonization sites, Landis argues that suitable candidates exist only around stars sufficiently similar to the Sun (in other words, single main-sequence stars within a restricted spectral range). Within a distance of 30 light years of Earth there are only five candidate stars, so a reasonable guess for this number is 5. These values produce a model that is close to critical: there are large colonized volumes of space and equally large empty volumes of space. According to the Landis model, then, we have not been visited by the many ETCs that exist in the Galaxy because we inhabit one of the voids.

*   *   *

The percolation approach addresses the Fermi paradox in an attractive way. Rather than attributing a uniformity of motive or circumstance to ETCs, it assumes civilizations will have a variety of drives, abilities and situations. The resolution of the paradox arises naturally as one possible consequence of the model. Of course, it is possible to argue about the details of the model; Landis himself does so in his paper. For example, the model ignores the peculiar motion of stars. Stars are not fixed, like the squares on a chessboard, but instead move relative to each other. Although the relative movement of stars is slow, it might affect the percolation model. It is also possible to suggest ways to improve the analysis. For example, we could develop more complex models, taking into account Galactic boundaries, habitable zones and the actual distribution of stars. One can also challenge the basic assumptions of the percolation approach. For example, is it realistic to assume the existence of a distance horizon, beyond which no civilization will ever colonize? After all, if a civilization can travel 50 light years, would a trip of 100 light years really be so much more difficult? And what of the assumption that only a few suitable stars will lie within the horizon? A suitably advanced civilization may well find it possible — indeed preferable — to construct habitats around a variety of stellar types. Furthermore, the Landis model assumes colonization will take place directly by members of an ETC. We will see in the next section that colonization might instead take place by probe — a process that is decidedly *not* described by a percolation model. If just one civilization successfully deployed probes to colonize the Galaxy, then the percolation model of Landis would fail.

Finally, even if this approach explains why we have not been *visited*, can it explain why we have not *heard* from an ETC? This question is particularly significant if one of the $p \geq p_c$ cases is true, and we inhabit a void surrounded on all sides by advanced civilizations. Even if daughter civilizations become independent of their parents, surely they would want to communicate with each other? Keeping in contact using radio or optical channels would be trivial compared to the problem of physically traveling between stars. It is hard to believe all these civilizations would travel, and then adopt and maintain a policy of silence. So why have we not overheard just one of these conversations? Why have we not seen a single "we are here" beacon? (In the Landis model, ETCs should have nothing to fear in revealing their position: one of the inputs to the model is that colonization of an inhabited system is so difficult it never takes place.) Why have we not seen one example of a massive engineering project, of the kind an advanced ETC might undertake? The answer to all these questions, of course, may simply be that we have not looked hard enough nor listened long enough. Nevertheless, although a percolation model provides an elegant explanation of why we have not been visited, I personally find it ultimately unconvincing.

SOLUTION 12: BRACEWELL–VON NEUMANN PROBES

> ...I looked to these very skies,
> And probing their immensities....
> Robert Browning, *Christmas Eve*

Interstellar travel is certainly difficult, perhaps impractical, but not impossible. Even with our present level of technology mankind has succeeded in launching a craft that will some day drift out to the stars. Imagine an ETC with a technology only slightly in advance of our own; suppose its craft travel at the sedate speed of, say, $c/40$. Then, if the ETC makes one more technological advance — the development of Bracewell–von Neumann probes — it possesses a strategy to colonize the Galaxy. And quickly.

*   *   *

Of the many contributions to science made by von Neumann (a partial list is on page 28), the most important may have been in the theory of computing. He became interested in computing at Los Alamos, where he was in charge of the calculations needed for the design of the bomb. Crude calculating machines had been developed to help von Neumann's team in its tasks; after the War, von Neumann turned his mind to what was required of more general-purpose computing machines. His considerations led to many of the important principles of computing, and most of today's

FIGURE 28   *John von Neumann (right) in conversation with Stanislaw Ulam (center) and Richard Feynman at Los Alamos.*

computers — which are based on the general logical design and mode of operation he championed — are known as von Neumann machines.[98]

The questions involved in the design of a general-purpose computing machine led von Neumann to ask an even bigger question: What is life? As a step toward answering this, he developed the idea of a *self-reproducing automaton*, a device that could (a) function in the world and (b) make copies of itself. (Such a device is also sometimes called a "von Neumann machine," but this leads to confusion with *the* von Neumann machine — the architecture that is at the heart of present-day computers. I will use the term "self-reproducing automaton" when I refer to this hypothetical device.) In von Neumann's scheme, the automaton has two logically distinct parts. First, it has a *constructor*, which manipulates matter in its environment to carry out tasks, including the construction of units it can then use to assemble a copy of itself. A universal constructor has the capacity to make *anything* — as long as it has suitable instructions. Second, it has a *program*, stored in some sort of memory bank, which contains the instructions needed by the constructor.

An automaton can reproduce itself as follows: The program first tells the constructor to make a copy of the program's instructions and place the copy in a holder. It then tells the constructor to make a copy of itself with a clear memory bank. Finally, it tells the constructor to move the copy of the program from the holder to the memory bank. The result is a reproduction of the original device; the reproduction can function in the same environment as the original and is itself capable of self-reproduction.

Of course, von Neumann did not give explicit details of how to *build* a self-reproducing automaton. Even today, we are far from being able to build such a device (although the seeming convergence of several technologies suggests that we may be able to do so in a few decades). What von Neumann was interested in was the logical underpinnings of self-reproducing systems, rather than any particular mechanism for achieving reproduction. In a lecture first given in 1948, he discussed the relevance of self-reproducing automata to the question of life. He argued that a living cell, when it reproduces, must follow the same basic operations as a self-reproducing automaton. Within living cells, there must be a constructor, and there must be a program. He was right. We now know that nucleic acids play the role of the program, and proteins play the role of the constructor. All of us are self-reproducing automata. (We discuss the function of nucleic acids and proteins later; see page 189.) What concerns us here is not what von Neumann's self-reproducing automata might tell us about life. Rather, it is how to use such automata to colonize the Galaxy. Frank Tipler outlined a possible scenario.

*   *   *

First, we must remember that the transport of intelligent beings to investigate planetary systems would be expensive: food, water, life support — all these items are necessary, but require energy to transport. Probes do not have this problem. Indeed, this is why Crick's motto for directed panspermia was "bacteria go further"; a small probe filled with a payload of bacteria would be cheaper to build and propel, and would enable an ETC to seed the Galaxy. With probes we are on the right track; but a bacteria-filled probe is of little use to an ETC wanting to explore and learn about the Galaxy. For an inquisitive ETC, it makes more sense to launch *Bracewell–von Neumann probes*. (These devices are usually called simply von Neumann probes in the literature. However, to the best of my knowledge, von Neumann never considered the possible uses of probes in interstellar exploration. The first person to suggest that probes would be useful for interstellar exploration and communication was Ronald Bracewell.[99] It seems reasonable, therefore, to refer to these devices as Bracewell–von Neumann probes.)

In Tipler's scenario, a Bracewell–von Neumann probe can be small: the payload need be nothing more than a self-reproducing automaton — one with a universal constructor and an intelligent program — and a basic propulsion system for use within the target system. After arriving at the target star, the program instructs the probe to find suitable material with which it can reproduce itself and make copies of the propulsion system. (If the planetary system resembled our own, then there would be plenty of raw material available for the constructor; asteroids, comets, planets and dust could all be broken down and utilized.) If necessary, radio signals from the home planet could send revisions to the program, so that the probe would never be out of date. Soon after arrival there would be a host of probes, each undertaking some pre-programmed task. Some might explore the planetary system, sending back scientific data to the home world. Some might construct a suitable habitat for later colonization by the home species. Some might even raise members of the original species from frozen embryos stored as part of the payload. And some would travel to another star, where the process would be repeated until every star in the Galaxy had been visited.

If probes traveled between stars at the rather stately speed of $c/40$, and if the propagation of the probes was directed rather than random, then a colonization wave could surge through the Galaxy in roughly 4 million years — a period that equates to just 2 hours 46 minutes of the Universal Year. This time is shorter than the colonization

FIGURE 29   *Ronald Bracewell has long been an advocate of SETI. He was also the first to suggest the use of probe technology as a means of exploring the Galaxy.*

times in the models of Newman and Sagan, and Fogg, but this is to be expected. Probes need not stay in a planetary system and wait for colonists to give them instructions on how to proceed: they already have their instructions. The Galactic colonization time is short because the process is *planned* to be efficient. Not only is colonization by probe quick, it is cheap. An ETC simply has to send the first few probes; after that, the Galaxy picks up the tab in terms of providing raw material for the continuing process.

Can such probes be built? Well, intelligent self-reproducing automata are certainly possible: Nature has already built them in the form of human beings. (As John Watson points out, humans are a good example of what we expect of a certain type of Bracewell–von Neumann probe! Perhaps we are not a "natural" species, but the probe technology of some advanced ETC?) Whether mankind can match Nature's accomplishment, or perhaps improve upon it and build better self-reproducing automata, is unknown. Certainly there are significant technical and engineering hurdles to overcome before we can build Bracewell–von Neumann probes. But even if mankind is not bright enough to develop probe technology, surely a technological civilization thousands or millions of years in advance of us could build probes. There seems to be no theoretical reason why they could not.

Colonization of the Galaxy by probe is technologically possible; it is quick; and it is cheap. Even if the aim is contact rather than colonization, Bracewell showed there are circumstances in which probes are more effective than radio signals. So as Fermi would ask: where are the probes?

We touched on this question in Chapter 3, when we discussed the possible use of probes in directed panspermia and when we considered places where a monitoring probe could hide. But such probes are not the Bracewell–von Neumann probes that can dismantle planets, undertake astroengineering projects, and colonize the Galaxy in the cosmological blink of an eye. There is no evidence of such probes ever having visited the Solar System, nor is there any evidence for their activity elsewhere in the Galaxy.

Even if an ETC has the ability to construct Bracewell–von Neumann probes, perhaps it would choose not to deploy the technology. There are risks, after all. The probes reproduce (like living beings) rather than replicate (like crystals), so inevitably there will be reproductive errors. There will be mutations. Probes would evolve, just as biological creatures evolve. The Galaxy could soon be home to different probe "species," each with its own interpretation of its goals. There would be a risk, for example, of a probe returning to the home system and failing to recognize it; not good news for the ETC if the probe's orders are to dismantle planets and use the material to construct something else. But is it a risk *every* ETC declines to take, and a problem *every* ETC fails to solve?

Since colonization of the Galaxy by probe seems straightforward, some authors argue there is a strong motivation for an ETC to engage in colonization: if species A does not do it, species B will. Stake your claim early, in other words. This sort of argument might have appealed to von Neumann, who was a strong proponent of the nuclear first strike. (In an interview with a *Time* magazine reporter, von Neumann said: "If you say why not bomb them tomorrow, I say, why not today? If you say five o'clock, I say at one o'clock.") We must be grateful that, in the 1950s and 1960s, wiser counsel than von Neumann's prevailed. Perhaps we can hope intelligent species develop to the stage where they have no urge to own every star, inhabit every planet, and populate the Galaxy with beings just like themselves. Nevertheless it takes only *one* ETC to reason that it should not take the risk of losing out on all that real estate.

*   *   *

A discussion of Bracewell–von Neumann probes is relevant to any discussion of the Fermi paradox, but you may ask why I present it in a part of the book devoted to *solutions* of the paradox. A surprising number of people seem to believe that probe technology *does* resolve the paradox. They argue that we do not see aliens because aliens would send probes rather than travel interstellar distances themselves. Of course, this entirely misses the point. Fermi's question refers either to aliens or the product of alien technology. After all, if we detected an object in space that was clearly artificial yet not man-made, then presumably we could deduce the existence of an extraterrestrial civilization that constructed the object. We see no evidence of aliens *nor of their probes*. Far from resolving the paradox, the possibility of Bracewell–von Neumann probes provides the paradox with real bite.

SOLUTION 13:   WE ARE SOLAR CHAUVINISTS

. . . the suns of home.
Rupert Brooke, The Soldier

We have implicitly assumed the important objects out in space are stable, middle-aged, G2-type stars like the Sun and watery planets like Earth. But who knows *where* a civilization much older than ours would choose to live? Earth-like conditions may be required for the genesis and early evolution of life, but once a civilization is technologically advanced and can construct a habitat for itself, it may not want to remain on the surface of a planet orbiting a commonplace star like the Sun. We tend to think ETCs would love to get their hands (or tentacles, or whatever) on the prime piece of real estate that is our Solar System, but that may simply be a reflection of our solar chauvinism. In which case the various Galactic colonization models may not be wrong; they may simply be inapplicable.[100]

For example, Dyson has suggested that a K2 civilization might choose to tear apart some of the planets in its system and use the material to create a sphere that encloses the star.[101] By doing this, all the star's energy output could be utilized; compare that with the situation on Earth, which intercepts only a billionth of the energy emitted by the Sun. If that civilization was also capable of interstellar travel, then presumably it could construct a Dyson sphere around any star that it visited. If so, why would it bother with our Sun, when so much more energy is available from stars of spectral class O? A star of spectral class O5, for example, pumps out

800,000 times more energy than the Sun. Perhaps, then, advanced ETCs are nomads, traveling from O-type star to O-type star in generation ships. They could arrive, enjoy a copious energy supply for the few million years of the star's life, then leave before the star goes supernova. The brilliant O-type stars provide unsuitable environments for life to *evolve*, because they die so quickly, but they might be the star of choice for K2 civilizations.

Alternatively, maybe advanced ETCs mine energy from the quantum vacuum or extract energy from black holes. In this case, would they require stars at all? They might live in their generation ships, never feeling the need to set foot (or alien pedal equivalent) on a planetary surface.

In short, perhaps the reason they have not been here is that there are many more attractive places to visit than we think. If this is the case, then the assumptions made in the various models of Galactic colonization are incomplete, and the conclusions may need to be revised.

SOLUTION 14: THEY STAY AT HOME...

There's no place like home.
J. H. Payne

One of the most thrilling events of my childhood happened on 20 July 1969.[102] My father woke me to watch Neil Armstrong and Buzz Aldrin land on the Moon. I guess most people of my age felt the same awe when they saw Apollo 11 touch down. More than thirty years later, we lack the ready capability — and motivation — to repeat the venture. Since Gene Cernan shook the lunar dust from his boots in 1972, no one has set foot on the Moon, and there are no definite plans for anyone to do so. Some space enthusiasts continue to do valuable work on establishing the factors needed for a manned trip to Mars, but such a trip is unlikely to happen soon. An assumption shared by many, including myself, is that intelligent species like ours will inevitably expand into space — so why are we not out there? Perhaps the assumption is wrong. Perhaps an unfortunate mixture of apathy and economics means ETCs stay at home; maybe that is the sad solution to the Fermi paradox.

There is reason to hope the suspension of manned space exploration is simply a pause. As technology improves, the journey into space will become cheaper and happen more frequently. We already have seen the first space vacationist, Dennis Tito, and more will surely follow him.[103] Indeed, the driving force behind manned space travel in the next few years may be tourism rather than science or high-tech industry.

In the longer run, there is a compelling reason why we should establish viable independent colonies on Mars or in O'Neill habitats: it would help ensure the survival of humanity should disaster strike Earth. In recent years we have come to understand what a dangerous world we inhabit. If a large meteor hit Earth we would be wiped out; if a super-volcano erupted, our technological civilization would crumble; climate change, whatever the cause, could destroy our way of life. Things have been peaceful here on Earth over the span of recorded human history, but our history corresponds to just 10 seconds of the Universal Year. Believing the world is calm because we have never seen it otherwise is like taking the attitude of a man who jumps off the top of a tall building and figures that, since 29 of the 30 floors have passed without incident, he is going to be okay.

In the even longer run, it makes sense to establish colonies around other stars in case something happens to the Sun. A coronal mass ejection only a few times more powerful than the most intense solar flare on record could cause us serious problems.[104] Ultimately, if we survive long enough, we will see the Sun moving off the main sequence on its way to becoming a red giant — and that really would force us to move home. (Zuckerman has shown that if the Galaxy contains between 10 and 100 long-lived civilizations, then almost certainly at least one of them would have been forced to migrate due to the death of its star.[105] If there are 100,000 such civilizations, then the Galaxy should have been completely colonized by civilizations whose home stars have evolved off the main sequence.)

Mankind has not exactly rushed headlong into space, but it is surely too early to say we will *never* attempt space travel. We have had the capacity to launch space vehicles for only a few decades; in the context of the Fermi paradox we have to think in terms of thousands or millions of years. And although it is probably fruitless to speculate upon the motives of putative extraterrestrials, there seems to be a universal logic to the establishment of off-world colonies. A species with all its eggs in one planetary basket risks becoming an omelette. Surely technologically advanced ETCs will move, however hesitantly, into space?

The idea that *all* ETCs stay at home seems (to me, at least) unlikely — unless there is a good reason why they should stay at home.

SOLUTION 15: ... AND SURF THE NET

> *Human kind*
> *Cannot bear very much reality.*
> T. S. Eliot, "Burnt Norton," *Four Quartets*

On page 51 we considered Baxter's suggestion that we exist in a virtual reality; the Universe appears devoid of life because advanced ETCs have engineered our reality to make it appear that way. We can invert the planetarium hypothesis to provide a less paranoid resolution to the Fermi paradox: maybe ETCs generate virtual realities for their *own* use. Maybe we do not hear from them because they stay at home and engage with an engineered reality more interesting and fulfilling than "real" reality.

It is easy to dream up scenarios in which an ETC might choose to disengage with the real world and instead inhabit a virtual one. For example, suppose their physicists discover a theory of everything, a goal our own physicists may be only a few decades from achieving. Suppose their biologists trace life back to its chemical origins and learn how to manipulate living material at the biochemical level. Their observational astronomers amass a wealth of data about the Universe, their theoreticians explain how the data fit their cosmological models, and their philosophers combine it all into a consilient theory of knowledge. In short, suppose they conclude that their science is finished. Furthermore, suppose the computing power available to this ETC is far in excess of our own: everything is wired, and their virtual reality simulations, which might feed in directly to their brains, provide satisfying sensory-rich experiences. Finally, what if such a civilization decided interstellar travel, although possible, is too difficult or costly to be worth the effort? Perhaps, under those circumstances, they would cease from exploration. They might instead investigate artificial realities.

We have no idea whether such a scenario is probable. For example, one could argue there will never be an end to the process of science; there will always be some new knowledge for a civilization to discover, new intellectual vistas to explore. But it is just as possible that the Universe obeys a small set of laws, and that the phenomena emerging from those laws are relatively small in number; in which case a long-lived technological society might eventually find that its science is essentially complete. (Although, of course, there is always art to consider as well as science.)

Similarly, one could argue that it is impossible to generate virtual realities as convincing as the reality we inhabit. Recall our discussion of the planetarium hypothesis, in which we considered the computing power required to generate a virtual reality sufficiently accurate to fool a civilization like our own. The computing demands were enormous, and the computing power required to fool an advanced civilization might be impossible to achieve. But the two cases are not equivalent. The computing power required to generate a virtual reality to satisfy *knowing* participants is much less than is required to *fool* mankind. In other words, the simulation designers could take shortcuts. There would be no need to calculate the trillions of interactions in a particle physics experiment; no need to simulate the outputs of protein-folding calculations; no need to present the results of gravitational microlensing observations. Their scientists would already have generated that knowledge in the "real" Universe. The simulation designers could thus instead concentrate upon generating satisfying and compelling simulations of objects and situations on the relatively restricted scale that intelligent beings (we believe) inhabit. This is not to say that the simulations need be restricted in imaginative scope: the situations to be simulated might be truly bizarre. But the participants of the virtual reality would not be "kicking the walls" of reality in the way that scientists and explorers do. All that is required is for the simulations to satisfy the participants. The requisite computing power is thus much less than is needed to create a full-scale Baxter planetarium.

My guess is that, if our own technology permitted it, a large fraction of humanity would prefer to live in a virtual reality. Already some people spend hours surfing the Net and prefer interaction with others to be mediated by computer. If simulations could provide us with a safe yet perfect sensory experience of walking on the surface of Mars, or hunting dinosaurs, or scoring the winning goal in a Cup Final, then I believe most of us would spend our time in those simulations. It would be infinitely better than TV — and consider how much time we waste on that.

The scenario of a stay-at-home surf-the-Net civilization seems to me to be an uncomfortably plausible future for mankind, but it does not alone solve the Fermi paradox. It is an example of a sociological condition that has to apply to *every* technological species for it to work. *We* may eventually prefer virtual reality, but why should couch-potatohood be a universal characteristic of intelligent species? Just as some of us prefer to interact with flesh-and-blood humans, so surely would at least some civilizations want to interact with others. Surely *some* ETCs would choose to explore, either directly or by probe.[106] Or, if interstellar travel proves impossible, would they at least not try to communicate?

Solution 16: They Are Signaling But We Do Not Know How To Listen

The world should listen then — as I am listening now!
Percy Bysshe Shelley, *To a Skylark*

Perhaps large-scale interstellar travel is unattainable, either for crewed starships or for probes. This would explain why we have not been *visited*, but not why we have not *heard* from them. Fermi simply asked: "Where *is* everybody?" The question refers to more than the mere absence of visitors; it refers to the absence of any evidence that they exist.

If interstellar travel is indeed unattainable — something ETCs would presumably quickly discover — then why should they hide? An ETC need not fear invasion by an aggressive neighbor, since any neighbors would be too distant to pose a threat. They have nothing to lose by signaling, and the potential reward is huge: mutually satisfying dialogs with equally advanced civilizations. Furthermore, telecommunication is cheaper than travel. (You are more likely to use the telephone or e-mail to keep in touch with antipodean relatives than you are to travel there by jet.) But if advanced civilizations are out there, educating each other, gossiping, holding conversations that are the Galactic equivalent of the Algonquin Round Table — then why do we not overhear them occasionally?

One extremely plausible answer is that we do not know how an ETC would choose to send a signal. We therefore do not know *how* to listen.

It is certainly true that we cannot know what communication technology ETCs might possess. As my editor pointed out, if a radio engineer from 1939 were somehow transported into the New York of 2002, he could build a radio receiver and conclude there were almost no useful radio broadcasts being made: he would not know about FM. Similarly, he would be blissfully unaware of communication devices employing lasers, fiber optics or geosynchronous satellites. So it is conceited of us to suppose we can know what communication channels are available to a technical culture that may be a million years in advance of our own. If they wanted to talk to each other secretly (maybe they do not want to influence the development of young species like our own?) then presumably they could maintain secrecy without difficulty. But things are very different if they *want* to be heard, and heard widely. We can assume that every civilization must obey the laws of physics; moreover, any ETC will know that other ETCs must obey those same laws. Since we all have to pay our energy bills, the number and types of signal that can reasonably be sent is quite restricted. Let us examine the advantages and disadvantages of four methods of communication: signals using electromagnetic waves, gravitational waves, particle beams, and hypothetical tachyon beams.

Electromagnetic Signals

The obvious way to send information is via electromagnetic (EM) radiation. Not only does it propagate at c, the fastest possible speed, it propagates over interstellar and intergalactic distances. (We know that EM signals can operate over interstellar distances because natural objects indicate their presence in this way over vast reaches of space. Astronomy is essentially the science of recording and interpreting these signals. We use visible light when we look at stars with our eyes or photograph them with optical telescopes. We use radio waves when we study the sky with radio telescopes. Increasingly we use the infrared, ultraviolet, X-ray and gamma-ray wavelengths, particularly in satellite experiments. If we can study natural objects over interstellar distances using the EM radiation they emit, then presumably we can do the same with artificial objects.)

For many years the working assumption made by researchers looking for ETCs is that technological civilizations will build powerful EM transmitters, broadcast a signal, and modulate it in order to convey useful information — perhaps, if we are lucky, they will broadcast their "Encyclopædia Galactica." In the next section I will discuss in detail how we might detect purposeful EM signals. Here, I want to argue that it may even be possible to detect EM radiation that leads to the discovery of *inadvertent* markers or beacons of K2 civilizations. (Detecting inadvertent markers of a K3 civilization might be even easier.) Even an inadvertent beacon would convey a tremendous amount of information: that life exists on another world, that it is technologically advanced, the location of that world, and so on.

We have already discussed why K2 civilizations might construct Dyson spheres. A Dyson sphere would radiate just as much energy as the central star — the energy has to go somewhere — but it would presumably do so in the infrared. In essence, the sphere would radiate because it is warm — about 200–300 K. So one way to search for an ETC would be to look for bright infrared sources at a wavelength of around 10 microns: such sources might be the waste heat from astroengineering projects.

A search by Japanese astronomers for artificial infrared sources out to a distance of 80 light years found no plausible signatures from Dyson spheres.[107] Although several stars show a large excess emission in the infrared, this happens to be because they are shrouded in dust. However, we cannot conclude from this that there are no ETCs within 80 light years; ETCs may choose not to construct Dyson spheres there for a variety of reasons. Even if Dyson spheres are common, *really* advanced civilizations — as Marvin Minsky pointed out[108] — would consider radiation at any temperature above the cosmic background temperature of 3 K to be wasteful. Perhaps an ETC advanced enough to construct a Dyson sphere is advanced enough to squeeze every last drop of useful work out of a star's radiation, leaving waste heat at 4 K. Perhaps we should be looking for points in space that possess a small temperature excess over the microwave background.

In 1980, Whitmire and Wright gave another example of how inadvertent beacons can be transmitted by electromagnetic radiation.[109] They asked what would happen if a civilization used fission reactors as an energy source over long periods of time. One of the problems with fission reactors is the need to dispose safely of radioactive waste material. And one proposed disposal method is to launch it into the Sun (though I, for one, would not be too thrilled at the prospect of having tons of radioactive waste perched on top of a chemical rocket). If an ETC used its star as a dumping ground for radioactive waste, then the spectrum of the star could exhibit characteristics that would not easily be interpreted as natural. For example, if we saw a stellar spectrum containing large amounts of the elements praseodymium and neodymium, then our interest would be caught. Furthermore, the alteration in the spectrum would not be a brief flicker; the spectral evidence of their nuclear waste disposal policy would be visible for billions of years. (A civilization might *deliberately* alter its star's spectrum in this way to create a beacon. This possibility was first suggested by Drake. Another method of using one's home star as a beacon was suggested by Philip Morrison: put a large cloud of small particles in orbit around the star in such a way that the cloud cuts off the starlight for a viewer who is in the plane of the cloud's orbit. Move the plane of the cloud and the distant viewer sees the star flash on and off. Variable stars naturally alter in brightness, but if the star flashed in a pattern that represented prime numbers, for example, then the distant viewer could quickly rule out a natural phenomenon.[110])

So far, no EM beacons — inadvertent or not — have been identified.

Gravitational Signals

Besides electromagnetism, the only other force we know that acts over astronomical distances is gravity. It too propagates at the speed of light, so perhaps ETCs might use gravitational waves to signal each other? Gravity, however, is a *much* weaker force than electromagnetism. To build a gravity-wave transmitter you have to be able to take large masses (of the order of a stellar mass) and shake them violently. It is debatable whether a K2 civilization would possess such technology. A K3 civilization might be able to build such a gravity wave transmitter, but why would it bother when EM waves do the job just as well and EM transmitters are so easy to construct?

FIGURE 30   *LIGO, in Washington State, consists of two 4-km arms at right angles, each with laser beams in high vacuum. There is an identical observatory in Louisiana, and the two installations will work in tandem. The objective will be to detect gravity waves by searching for changes in length a thousand times smaller than an atomic nucleus.*

The complementary problem of detecting gravitational waves is also much more difficult than the equivalent problem of detecting EM waves. It is so difficult, in fact, that terrestrial science has yet to build a functioning gravitational-wave detector. (Detectors such as LIGO will soon come online, but even if they are successful they will have the sensitivity to detect gravitational waves from only the most violent astronomical phenomena.[111] The detectors will collect exceptionally interesting scientific data, but they will not find modulated signals.) So, given the difficulties of transmitting and receiving gravitational waves, it seems unlikely that an ETC would use them for communication.

Particle Signals

Cosmic rays, in the form of electrons, protons and atomic nuclei, can reach Earth over interstellar distances — and cosmic-ray astronomy is a thriving research field. However, charged particles like these would constitute a poor choice of communication channel because a transmitting civilization could not guarantee where the particles would end up: twisting magnetic fields throughout the Galaxy make the paths of these particles quite tortuous. Neutrinos are electrically neutral, so at first glance they seem a better choice for a communication channel. Unfortunately, neutrinos are difficult to study because they react so infrequently with matter; typically, a neutrino will pass through 1000 light years of lead before stopping! Nevertheless, despite the tremendous difficulties involved, astronomers have developed neutrino telescopes. [112]

Neutrino Telescopes

The first such telescope was developed by Ray Davis, who wanted to study the neutrinos that are generated in nuclear fusion reactions in the heart of the Sun. His telescope was in essence a 100,000-gallon vat of perchloroethylene (dry-cleaning fluid) buried almost a mile beneath the ground in the Homestake gold mine in South Dakota. It was the strangest telescope anyone had ever constructed (there are stranger telescopes nowadays), but the setup was necessary because neutrinos are so elusive. The deep mine shielded the vat from other subatomic particles that bombard Earth; the dry-cleaning fluid provided enough chlorine atoms to guarantee detectable numbers of neutrinos.

Theory predicted that when a chlorine nucleus captured a neutrino it would turn into a nucleus of radioactive argon. So by detecting argon atoms, Davis was able to detect solar neutrinos. Of 10^{21} neutrinos passing through the vat each day, theory suggested that 6 events should take place; but the experiment found only 2 events per day. Davis' experiment continues to detect solar neutrinos, but only one third of the expected number — a finding that is of great significance for particle physics.

FIGURE 31 *A virtual reality deep view of the 0.1-km² Antares neutrino telescope, which will be located beneath the Mediterranean. Similar detectors are situated in mine shafts and underneath mountains.*

In February 1987, the Kamiokande detector in Japan and the IMB detector in America between them stopped 20 neutrinos in a period of a few seconds. Those neutrinos were produced in the famous supernova of that month: SN1987A. Now, SN1987A occurred in the Large Magellanic Cloud, about 170,000 light years away. Demonstrably, then, it is possible for neutrinos to travel interstellar, even intergalactic, distances and for a primitive technological civilization like ours to detect them. Perhaps ETCs use modulated neutrino beams to communicate with each other? Well, perhaps. But again we have to ask why they would do this when electromagnetic waves do the job far better and much more cheaply.

Tachyon Signals

We can speculate that extremely advanced ETCs will use tachyons to signal each other. If tachyons exist, and if it is possible to modulate a beam of them to carry signals, then no doubt they will be an attractive option for interstellar communication. Tachyon-based communication would obviate that irritating delay between asking a question and receiving an answer — a delay that can be hundreds or thousands of years. Unfortunately, as we saw earlier (see page 67), there is absolutely no evidence that tachyons exist, let alone that it is possible to use them to send signals.

*   *   *

Perhaps there are lots of civilizations out there, communicating with each other using gravitational waves, neutrinos and tachyons. Or perhaps they send signals using techniques we have not yet dreamed of — techniques that break no laws of physics but that are as exotic to us as fiber-optic communication channels would be to a 1939 radio engineer. Since we cannot detect such signals it would explain why we have not heard from them; it would explain the "great silence" — if not the full Fermi paradox itself.

On the other hand, even for advanced civilizations, communication by EM waves seems to be a logical choice: EM signals are cheap to produce, the message moves as fast as is possible in a relativistic Universe, and the signals are easy to receive. If an ETC *wanted* to make its presence known to other perhaps less developed civilizations (civilizations like us, who *can* only listen for electromagnetic signals), then the EM spectrum might be their only option.

For these reasons, although it may seem conceited and it may mean we are missing out on Galactic conversations, many physicists would argue that we know how to listen for signs of extraterrestrial civilization: we should listen for their EM radiation. (In fact, given the level of our present technology, we have little option but to try and detect such radiation.) But at what frequency should we listen?

SOLUTION 17: THEY ARE SIGNALING BUT WE DO NOT KNOW AT WHICH FREQUENCY TO LISTEN

57 channels and nothing on.
Bruce Springsteen

If ETCs do indeed use EM radiation to communicate with each other or to notify their presence to less advanced civilizations, then there are several different types of signal we might search for.

The easiest type of signal to detect would be one that an ETC has deliberately targeted at us. It is not too arrogant of us to suppose a nearby ETC would beam signals toward the Sun. Advanced civilizations would classify the Sun as a good candidate for possessing life-bearing planets, and they could probably detect the existence of Earth over interstellar distances. With our present level of technology we can detect Saturn-sized planets around other stars, so advanced ETCs will be able to do much better. If they beam signals to target stars in the hope of making contact, then our Sun would be on their list. (Upon re-reading this paragraph, some of the statements sound too definite. We are in the realm here of trying to second-guess the motives and intentions of putative aliens — an enterprise fraught with risks. But we have to begin somewhere.)

A second type of signal would be one meant for communication but targeted elsewhere, a signal we might nevertheless overhear. Yet another type of signal would be one not intended for communication at all, but instead leakage from other activities — just as EM signals leak out from Earth due to our radio and television transmissions, and our use of military radar. (Such signals have been leaking from Earth for several decades, but developments in cable and satellite telecommunications systems suggest they may soon cease. Perhaps the same will be true for ETCs, and the period over which a technological civilization is "radio-bright" can be measured in decades, in which case we have essentially no chance of discovering this type of signal. On the other hand, maybe future technological developments — solar satellites that beam energy back to the home planet in the form of microwaves, perhaps, or navigational beacons for steering through a crowded planetary system — would leak EM radiation into space.)

With our present level of technology, it makes little sense to look for leakage radiation. We should do the easy things before attempting harder projects, and it is easier to detect radiation intended for communication. But at which wavelength will ETCs choose to transmit? In other words: at what frequency should we listen?

*   *   *

The EM spectrum is extremely broad. Visible light, which reaches from 7.5×10^{14} Hz (deep violet) to 4.3×10^{14} Hz (red) forms a minuscule part of the spectrum. Ultraviolet, X-rays and gamma-rays have progressively higher frequencies, reaching up to 3×10^{19} Hz or higher. Infrared, microwaves and radio waves have progressively lower frequencies, reaching down to 10^8 Hz. Our technology employs all these wavelengths for a variety of purposes, ranging from medical applications (X-ray frequencies) to household devices (garage door openers work at 40 MHz, for example, and baby monitors at 49 MHz). There seems to be a frequency for everything. So which frequency is best for interstellar communication?

In the late 1950s, Philip Morrison and his colleague Giuseppe Cocconi were among the first to consider this question. Astronomers had by then developed radio telescopes and were using them to make significant discoveries. It was against this background that Morrison investigated the possibility of using gamma-rays as a different window on the Universe. As part of this work he showed how gamma-rays, unlike visible starlight, could travel across the dusty plane of the Galaxy. He told Cocconi of this result, and his colleague pointed out that particle physicists already generated gamma-ray beams in their synchrotrons; why not send the beam into space and see if an ETC could detect it? It was a fascinating question, and it got Morrison to thinking about the prospects for interstellar communication. He replied that they should consider not just gamma-rays but the whole EM spectrum — from radio waves all the way up to gamma-rays — and choose the most effective band for signaling.

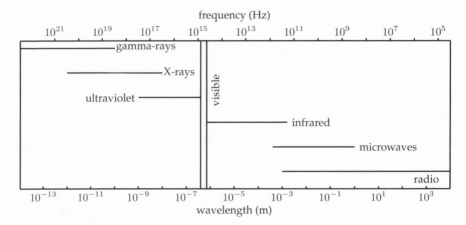

FIGURE 32 *The wavelengths and frequencies of the electromagnetic spectrum. The horizontal lines appear on a logarithmic scale: each "tick" corresponds to a factor of ten. It is clear from this diagram that visible light corresponds to only a small fraction of the electromagnetic spectrum.*

They quickly concluded that visible light was a poor choice for signaling, since the signals would have to compete with starlight; gamma-ray telescopes were not feasible at that time; the radio band seemed the best bet. The Arecibo radio dish in Puerto Rico was the appropriate instrument with which to search for signals: they calculated that if an ETC had its own Arecibo and used it to transmit a directed beam at a sharply tuned frequency, then our Arecibo could detect the alien dish from halfway across the Galaxy.[113]

Narrowing down the search to the radio band was a major advance, but it still left many possible frequencies. Radio waves can be anywhere between about 1 MHz and about 300 GHz.[114] This is bad news, for the following reason. If an ETC wishes to transmit a signal over large distances, then it needs to send a *narrowband signal* — a signal at a precise frequency — since broadband signals are too easily mistaken for background noise. (When you twiddle the dial on a radio, you hear the background hiss of broadband noise between the narrowband signals of the radio stations.) The narrowest frequency generated naturally is by an interstellar maser. (Masers, which amplify microwaves, act very much like lasers.) It has a width of about 300 Hz; anything much narrower than this is a candidate for an artificial signal. Suppose, then, that ETCs transmit signals with a bandwidth of 0.1 Hz. (It makes little sense to transmit over interstellar distances with a bandwidth less than 0.1 Hz, since electrons in interstellar clouds will tend to disperse the signal.) This means that we have a *huge* number of radio channels to search through. Unless we narrow the search even further (or we get extremely lucky) we could be searching for a long time.

FIGURE 33   *The Arecibo radio telescope, located in Puerto Rico, is a huge structure. The dish itself is 305 m in diameter, 51 m deep, and covers an area of about 8 hectares. This telescope could detect an alien transmission from the other side of the Galaxy.*

Cocconi and Morrison pointed out that at frequencies less than about 1 GHz the Galaxy is noisy. It makes little sense to send a signal at a frequency lower than 1 GHz because background noise would drown it. On the other hand, at frequencies higher than about 30 GHz Earth's atmosphere becomes noisy. If an ETC were broadcasting at frequencies higher than 30 GHz, we would be unlikely to detect the signal because of atmospheric interference. In fact, the quietest region is between about 1 GHz and 10 GHz. Cocconi and Morrison suggested that it makes most sense to search for radio signals in that region, where an artificial signal would really stand out.

They refined the frequency range even further. They pointed out that clouds of neutral hydrogen — the simplest and most common element in the Universe — strongly emit radiation at 1.42 GHz. Every intelligent observer in the Universe will know about the hydrogen line. It makes sense to look there. Soon after, it was discovered that the hydroxyl radical radiates prominently at 1.64 GHz. Hydrogen, H, and hydroxyl, OH, together make up the compound water: HOH — or H_2O. Now water, as far as we know, is absolutely necessary for the existence of life. Find water, and you have a chance of finding life. And since the region between 1.42 and 1.64 GHz is about the quietest part of the radio spectrum, it seems a logical place for a civilization to broadcast if it wants to attract attention. This band has been dubbed the *waterhole*. It is a beautiful name, conjuring up visions of many different species coming together at a life-giving source of water.

*   *   *

At about the same time that Cocconi and Morrison presented theoretical reasons for listening in the long-wavelength region near the hydrogen line, Frank Drake was *doing* exactly that: listening for signals near the hydrogen line. Drake had built equipment to study this part of the radio spectrum for mainstream astronomical purposes, but he had an abiding interest in the possibility of extraterrestrial life. He used the radio telescope at Green Bank to listen to two stars — Tau Ceti and Epsilon Eridani — for signals. His Project Ozma was the first time mankind had searched for an ETC. Although the results were negative, Drake's observations — along with the Cocconi–Morrison paper — proved to be a watershed for SETI.

FIGURE 34   *Frank Drake is a towering figure in the SETI field. In addition to the eponymous Drake equation, he is known for carrying out the first radio search for an ETC.*

The situation now seems much more complicated than it did four decades ago for Drake, Cocconi, and Morrison. They knew only of one spectral line, the hydrogen line, so the choice of where to search seemed quite clear. Modern astronomers, however, are aware of tens of thousands of spectral lines emanating from more than 100 types of molecule in interstellar space. There are very good arguments why we should study other frequencies. (Important examples include 22.2 GHz, which corresponds to a transition of the water molecule, and simple multiples of the hydrogen-line frequency — twice the hydrogen-line frequency, π times the hydrogen-line frequency, and so on. There is a particularly attractive "natural" frequency for intergalactic communication, which I discuss in a later section.)[115] Although many authors maintain that the waterhole is the "natural" place to search for signals from within our Galaxy, we may eventually be forced to search through the whole window from 1 to 30 GHz.

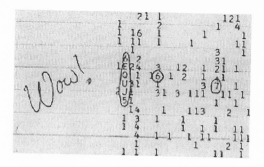

FIGURE 35 *The famous "Wow" signal. The Ohio State University Big Ear Observatory scanned 50 channels and recorded the observations on a printout sheet. For each channel a list of letters and numbers appeared on the printout. In the Big Ear system, numerals 1 to 9 represented a signal level above background noise. For strong signals, letters were used (with Z being stronger than A). On the night of 15 August 1977, Jerry Ehman spotted the characters "6EQUJ5" on channel 2. This signal started from roughly background level, rose to level U, then decreased back to background level in 37 seconds. This was exactly what an extraterrestrial signal might look like; Ehman circled the characters and wrote "Wow!" in the margin.*

In over 40 years of listening, none of the radio searches has found an extraterrestrial signal that is clearly artificial in origin. That is not to say that no signals have been found, of course. (Drake himself detected a signal emanating from the general direction of Epsilon Eridani, just a few hours after the commencement of Project Ozma. However, further investigation showed that the signal was clearly terrestrial in origin.) The radio searches have detected lots of signals, many of them rather intriguing. The famous "Wow!" signal is typical of the best signals found so far. It was a powerful narrowband spike, with characteristics indicating that it almost certainly came from space, but when Big Ear listened again to that part of the sky the signal had gone. Several attempts to relocate the "Wow!" signal have failed. Recently, for example, searches with the Very Large Array enabled astronomers to investigate two hypotheses regarding the signal. First, perhaps it came from a weak yet steady transmission, which momentarily increased in strength due to scintillation (like the twinkling of a star). Second, perhaps the signal was a powerful pulse, designed to attract attention to a much weaker continuous signal. Both possibilities seem to have been eliminated. Nothing interesting was found, down to a level that was 1000 times weaker than the original signal.

The "Wow!" signal *may* have emanated from a distant civilization, a beam that happened to sweep across Earth's path one August night and then moved on. But it seems much more likely that the signal came from a man-made satellite.[116]

SETI Projects

Since Project Ozma there have been more than 60 SETI projects, most of which have searched the waterhole region. In recent years, the projects have become increasingly sophisticated. Project META (Million-channel Extra-Terrestrial Array), developed in 1985 by Paul Horowitz,[117] could study a million channels at once in the waterhole region. In 1990, META II started searching the southern sky, monitoring 8 million extremely narrow 0.05-Hz channels near the hydrogen line at 1.42 GHz, and also at twice this frequency, 2.84 GHz. In 1995, Horowitz initiated Project BETA (Billion-channel Extra-Terrestrial Array), which scans the waterhole region at a resolution of 0.5 Hz. From META to BETA in just ten years is significant progress! Project SERENDIP (Search for Extraterrestrial Radio Emissions from Nearby Developed Intelligent Populations) piggybacks on radio telescopes being used for other astronomical purposes. The drawback with this approach is that there is no choice over where to listen; it can look for signals only where the telescope happens to be pointing. On the other hand, since it does not interfere with the normal functioning of the telescope, the project can be run continuously.[118] The present incarnation of the project piggybacks on the Arecibo telescope and searches 168 million channels, each 0.6 Hz wide, near 1.42 GHz. Southern SERENDIP piggybacks on the Parkes Observatory in Australia to search the southern sky, also at the hydrogen line. Project Phoenix, which began in February 1995, is halfway through a search for signals within the range of 1.2 to 3.0 GHz in channels that are just 0.7 Hz wide.

Despite the increasing sophistication of radio SETI, sorting through billions of channels in the hope of finding a signal remains a laborious task. Is there really no alternative to the microwave/radio part of the electromagnetic spectrum? It happens that there is.

At about the same time that Cocconi and Morrison suggested listening for radio transmissions, Arthur Schawlow and Charles Townes outlined the working principles of lasers. Early devices were feeble, but just as computing power has increased geometrically, so has the power of lasers. It now seems clear an advanced ETC could communicate its presence using laser pulses and might prefer this method over radio. Not only would a short pulse of laser light stand out even over interstellar distances, it would plainly be artificial. Furthermore, an ETC could send beacon signals to millions of stars each day. Perhaps we should not be listening for radio signals alone; we should also be looking for signals in the visible spectrum.[119]

FIGURE 36   The Very Large Array in Socorro, New Mexico. The array consists of 27 dishes, each of which is 25 m in diameter. Despite their appearance in the movie Contact, *the telescope only rarely listens for broadcasts from extraterrestrials. Recently, though, it tried to relocate the "Wow!" signal — but unfortunately found nothing unusual.*

Optical SETI is not as advanced as traditional radio SETI, but this is changing thanks mainly to the efforts of Stuart Kingsley. Kingsley uses his COSETI (Columbus Optical SETI) Observatory to look for narrowband laser signals from a list of target stars. It is encouraging that the equipment required for such a search is relatively simple and within the range of the dedicated amateur astronomer.[120] Professional SETI scientists have caught on, however, and are beginning to develop large-scale projects.[121]

Even gamma-rays have been suggested as a communications channel for civilizations in contact over intergalactic distances. John Ball hypothesizes that gamma-ray bursters are messages sent by ETCs. However, although the detailed origin of these events is still being debated, nearly all astronomers believe bursters are a natural phenomenon. We have to employ Occam's razor once more: if we can explain bursters as a natural phenomenon, then Ball's hypothesis is simply unnecessary.

*   *   *

In 40 years of searching — mainly in the radio, but occasionally in the infrared and increasingly in the visible — astronomers have detected no signal. To rephrase Fermi's question: where *are* the signals? The lack of signals means that we can now start to place limits on the number and type of ETCs in our neighborhood. Some authors claim that this null result means that we can rule out the presence of K2 and K3 civilizations not only in our Galaxy, but beyond even our Local Group of galaxies.[122] This claim is overstated, since it rests on several assumptions that may not be valid. Nevertheless, taking a conservative viewpoint, we can probably rule out the existence of a K3 civilization anywhere in our Galaxy and a K2 civilization in our particular part of the Galaxy: if they were there, we would surely have heard from them. In a few years, if the null result continues, we will be able rule out the existence of K1 civilizations out to 100 light years.

Billions of channels and — so far — nothing on.

SOLUTION 18:   OUR SEARCH STRATEGY IS WRONG

We seek him here, we seek him there.
Baroness Orczy, *The Scarlet Pimpernel*

Even if ETCs are broadcasting radio signals, and we are tuned to the correct channels, where should we point our telescopes? The sky is large, and our resources are few. It would be tragic to train our telescopes on Canopus, say, if the civilization on Capella were trying to catch our attention.

We can employ two search strategies. A *targeted search* focuses upon individual nearby stars. It uses instruments of great sensitivity in the hope of detecting signals deliberately beamed toward us or leakage radiation that happens to pass our way. A *wide-sky survey* scans large areas of the celestial sphere and thus encompasses a myriad of stars. The sensitivity of a wide-sky survey is vastly inferior to a targeted search.

The earliest search for an ETC — Drake's Project Ozma — targeted just two stars: Tau Ceti and Epsilon Eridani. Of modern targeted searches, the best known is Project Phoenix: it targets a list of about a thousand old, Sun-like stars within a distance of 200 light years, and listens for signals within the range 1.2 to 3.0 GHz in channels just 0.7 Hz wide — so for each star more than 2.5 billion channels are checked. However, most of the large SETI projects currently in operation — such as SERENDIP, Southern SERENDIP and BETA — are wide-sky surveys. Future projects — such as the SETI League's plan to link the observations of 5000 small radio telescopes — will be wide-sky surveys.[123] Targeted searches are a rarity; of today's major radio searches, only Project Phoenix employs a targeted strategy. Maybe we are employing our precious SETI resources in the wrong manner? Maybe we do not see ETCs because we are not searching with sufficient sensitivity? Should we not look hard and long and deep at planetary systems that might harbor life, instead of skimming across the sky?

Well, no. It turns out that the modern wide-sky surveys are doing the right thing. An analysis by Nathan Cohen and Robert Hohlfeld shows that we should play the numbers and look at as many stars as possible.[124]

In Nature, we often find that objects with a large value of some property are rare, while objects with a smaller value of that property are common. Thus, bright stars of spectral class O are few in number, while dim M-class stars are widespread. Strong radio sources like quasars are rare, while weak radio sources like stellar coronas are common. Which are we more likely to detect: the rare "bright" objects or the common "dim" objects? It depends on the strength of the rare sources compared to the common sources. For example, quasars are *incredibly* strong radio emitters; it does not matter that they are at extreme distances — they far outshine the closer but weaker stellar sources. Thus, radio telescopes in the early 1960s could detect rare, distant quasars more readily than common, nearby sources. In the same way, even if advanced ETCs are incredibly rare, Cohen and Hohlfeld showed that we are more likely to detect their beacons than weak signals from a host of ETCs not much more advanced than ourselves. (The only way to avoid this conclusion is if the stars are teeming with intelligent life. If ETCs are common, then a targeted search like Project Phoenix is likely to find one in its list of target stars.) Wide-sky surveys are therefore more likely to produce positive results; at the very least, when we pick targets for in-depth study, we should try to ensure that the receiving beam contains galaxies or large clusters of stars behind the target.

A Frequency for Intergalactic Communication

A "natural" frequency for intergalactic communication is represented by $f = \frac{k}{h}T_0 \approx 56.8\,\mathrm{GHz}$, where T_0 is the observed temperature of the cosmic background radiation, k is the Boltzmann constant, and h is the Planck constant (it thus links the regimes of cosmology and quantum physics). This frequency was originally proposed in 1973 by Drake and Sagan, and independently by Gott in 1982.

I have a tiny feeling of unease with the wide-sky surveys, and this harks back to the problem of the frequency at which we should listen. The surveys take in distant galaxies, and most surveys listen at or around the waterhole. But there is a better frequency than the waterhole for intergalactic (as opposed to interstellar) communication: 56.8 GHz. This frequency is tied to the observed cosmic microwave background, so it is a universal frequency. If an ETC in a distant high-redshift galaxy emitted a signal at a frequency related to the above, then it could be sure the signal might be received at any future time. The signal could potentially reach large numbers of galaxies.[125] (There is another factor to consider here. On Earth it took about 4.5 billion years for a technological civilization to arise. If this is the time it takes other civilizations to arise, then — depending upon the exact details of the cosmological model one prefers — it is pointless looking at galaxies with redshifts much larger than 1. The light we now see from these distant galaxies set off when the Universe was only about 4.5 billion years old; there would not have been time for a K3 civilization to arise.) Unfortunately, Earth's atmosphere has a wide oxygen absorption band at 60 GHz, which means our radio telescopes cannot carry out a search at 56.8 GHz. Observations at this frequency will have to be performed from space. In the meantime, perhaps a K3 civilization in a faraway galaxy is signaling us right now.

*   *   *

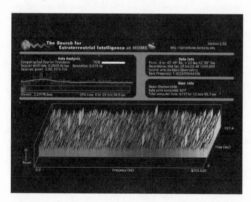

FIGURE 37   *A black-and-white screenshot of the SETI@home screensaver.*

I cannot leave this discussion without mentioning one of the most innovative of recent scientific projects. Since Drake first pointed his radio telescope to Tau Ceti in the hope of finding a signal, engineers have improved the sensitivity of radio receivers by a factor of about 20, and astronomers have amassed much more knowledge about the birth and evolution of stellar systems. But the biggest development since the Project Ozma days has been the remarkable increase in available computing power. The SETI@home project, founded by David Gedye, has harnessed this power in a way that has captured the enthusiasm of the general public as perhaps no other scientific project has done.[126] Participants download a small client program for their home or work computer. The program usually works as a screensaver; in essence, when the user's computer is not engaged in "proper" work, the client program comes to life and begins calculations on a packet of data — known as a work unit — taken by the Arecibo radio telescope. Once the calculations are complete, the program sends the work unit back to SETI@home, where it is merged with all the other results from around the world, and a new work unit is downloaded. More than a million CPUs have crunched data from Arecibo, and they have combined to make SETI@home the world's largest and most powerful virtual computer.[127] This immense computing power has enabled astronomers to make one of the most finely tuned searches for ETCs ever attempted: the program looks at data from a band with a width of 2.5 MHz centered at the 1420 MHz hydrogen line, and examines channels as narrow as 0.07 Hz.

*   *   *

New projects like SETI@home — and traditional projects like SERENDIP and BETA — seem to have got the search strategy right: look at wide areas of the sky, across billions of stars, and hope that somewhere in that vast collection we find a very rare yet very powerful transmission.

So far, we have heard nothing.

SOLUTION 19:   THE SIGNAL IS ALREADY THERE IN THE DATA

I do not search; I find.
Pablo Picasso

Forty years of SETI projects have amassed a huge amount of data. Is it possible that somewhere in all that data there is a thumbprint of an ETC — a signal we have not yet recognized?

The SETI detectors can be fooled by a host of terrestrial signals: stray radiation from mobile phones, radar from military devices, and so on. The SETI astronomers are alert to these sources of interference and can usually identify them for what they are. But there remain a few tantalizing exceptions. For example, the META project logged several signals that were non-random and possibly intelligent transmissions.[128] Zuckerman and Palmer examined 700 nearby stars and logged ten signals that could have been artificial.[129] We have already discussed the famous "Wow!" signal.

The trouble is, whenever astronomers redirect their telescopes in the direction from whence the signal came, they find nothing. The signals never repeat. Perhaps these signals were indeed the intermittent broadcasts of ETCs, a lighthouse beam that swept across Earth before moving away. Or perhaps they were simply a source of radio interference that has not yet been identified.

105

Another problem arises with the interpretation of the data from telescopes. We collect photons from gamma-ray bursters and explain their origin in terms of a cataclysmic fireball; we collect photons from stars with an infrared excess and deduce that the star is shrouded in dust; we find a thermal spectrum and infer that it comes from a black body. We could explain all these observations in terms of ETC activity. As we have seen, Ball suggested that ETCs might communicate by exchanging bursts of gamma-rays; one of the signatures of a Dyson sphere is an infrared excess; the most efficient mode of communication (which an ETC would presumably employ) is indistinguishable from black-body radiation to observers like us, who are not privy to the system used.

Ultimately, the difficulty is that we are stuck on a tiny rock, at the bottom of a thick atmosphere, trying to make sense of the Universe by interpreting the occasional photons our telescopes can catch. This is a challenge. Sometimes scientists may be wrong; but if we *can* explain observations in terms of natural phenomena, then we need not postulate the existence of ETCs. Occam, again. So when we observe, for example, that the spectra of almost all galaxies show a redshift, it is enough to explain it in terms of the expansion of the Universe — an explanation fantastic (and beautiful) enough in itself. We do not need to suppose, as did one SF story, that redshifts are the exhaust gases of alien craft fleeing from mankind.

We have to hope advanced ETCs will make their signals unambiguous and clearly distinguishable from noise. We have to hope their signals will be strong; if our present generation of detectors is insufficiently sensitive for the task, then 40 years of observation will have been wasted. And we have to hope they repeat their signals often. It would be a pity if we have already recorded a signal but cannot prove it is from an ETC.

SOLUTION 20: WE HAVE NOT LISTENED LONG ENOUGH

> *Patience is bitter, but its fruit is sweet.*
> Jean-Jacques Rousseau, *Emile*

In 1991, Drake wrote about his hopes for detecting signals from an ETC: "This discovery, which I fully expect to witness before the year 2000, will profoundly change the world."[130] Ten years on, much has happened in SETI research. The field is thriving. But the discovery has not been made. Perhaps Drake was simply being impatient. Perhaps the answer to the Fermi paradox is that ETCs are out there, communicating with each other and maybe even attempting to communicate with us, but that we simply have not listened long enough for our search to bear fruit.

This is the position SETI enthusiasts take, and for good reason. Consider, for example, some of the difficulties the Arecibo telescope has in receiving a signal from an ETC. One is that the Arecibo receiving beam area covers only a small patch of sky at any particular time, so there are millions of slightly different directions in which astronomers can point the telescope. Another is that for each patch of sky there are billions of frequencies to check. Yet another difficulty is that a signal might take the form of a burst rather than a continuous beacon; to detect a burst, Arecibo has to be pointing there at the right time. In short, to detect a radio signal from an ETC, our telescopes must be pointing in the right direction at the right time and tuned to the right frequency. There are trillions of possible combinations of these parameters, of which we have checked only a fraction. If ETCs chose to chatter at each other using *lasers*, then it is extremely unlikely that Earth would be in the path of any of the beams; billions of civilizations could be out there, talking to each other, and we would not hear them. It seems not unreasonable, then, to say we have not searched long enough. Perhaps we simply have to be patient.[131]

Some people, however, believe this to be an unsatisfactory resolution of the Fermi paradox. In a sense, the crux of the paradox is that we have been "waiting" for evidence of extraterrestrials for billions of years: they themselves, or their probes, or at least their signals, *should already be here.* Evidence of their existence, whatever form such evidence might take, should have been here long before mankind began to wonder if other species were out there. Spending a few more decades observing, with admittedly much more powerful technology, is missing the point.

Let us consider it another way. How many ETCs presently inhabit the Galaxy? Sagan and Drake suggested there might be 10^6 ETCs in our Galaxy at or beyond our present level of technological development (so on average there should be an ETC within 300 light years of Earth).[132] A more conservative estimate by Horowitz is there might be 10^3 advanced ETCs in our Galaxy (so, if they are randomly distributed through space, there will be an ETC within 1000 light years of Earth). If these 10^3 to 10^6 civilizations are long-lived — perhaps billions of years old — then they must surely have a Clarke-level of technology (one that, to us, is indistinguishable from magic). Even if they do not want to travel, or find it impossible to travel, surely such civilizations could make it easy for us to see them; why don't they? Alternatively, these civilizations might be short-lived. (Many authors often set parameters in the Drake equation in such a way they arrive at the relation $N = L$. In other words, the number of civilizations out there right now is equal to their average lifespan.) If there are 1000 civilizations now, and if the rate of formation of technological civilizations has been more or less constant over the history of the Galaxy, then about 10 billion civilizations will have lived and died in our Galaxy alone. Is it likely that not one ETC left any record of their hopes, their achievements, their existence? (If true, it is an almost unbearably sad thought.)

We return to the question: where are they — either their craft or their probes or their signals? We should not have to *wait* for evidence of their existence — the evidence should already be here.

SOLUTION 21:   EVERYONE IS LISTENING, NO ONE IS TRANSMITTING

Never the least stir made the listeners.
Walter de la Mare, *The Listeners*

We have briefly discussed the difficulties in trying to *receive* a signal from ETCs. We have not considered how difficult it might be for them to *send* a signal. One thing seems certain: no matter how hard it is to detect a signal from an unspecified planetary system among the Galaxy's hundreds of billions of stars, it must be a lot harder to send it — at least, to send it with any expectation that it will be detected. Could it be everyone is listening and no one is transmitting?

In a sense, our civilization already transmits signals to the heavens. For several decades, our radio and TV transmitters have been leaking EM radiation into space. As I write, live broadcasts about the fall of the Berlin Wall could be sweeping across the star Tau Ceti; news of Kennedy's assassination could now be reaching Arcturus; cricket lovers in the Castor system may soon receive word of Bradman's last Test innings. But it is debatable whether these transmissions will be detected, even if ETCs are listening. Our transmitters direct their beams horizontally, to be picked up by individual antennae. So although some of the output is lost to space — a beam of EM radiation sweeps across space as Earth rotates on its axis and as it orbits the Sun — it is down to luck whether any of it intersects with a distant star. Furthermore, the high bandwidth and relatively low power of our transmitters mean even an Arecibo-type telescope would struggle to detect our broadcasts much beyond the orbit of Pluto. So unless ETCs are nearby, extremely lucky, and have a level of receiving technology far beyond our own, they are unlikely to detect our inadvertent transmissions.[133] Besides, the amount of this leakage radiation is lessening as we increase our use of cable. (The radiation from powerful military radars, and the signals that astronomers bounce off Venus and Mars to map the topography of those planets, has more chance of being detected over interstellar distances. On the other hand, such radiation is highly focused; the beam is unlikely to intersect with an alien receiver.)

What if we *wanted* to be noticed? Rather than trusting to luck and hoping an ETC spots our TV (hoping too, perhaps, they receive *Cheers* rather than *Charlie's Angels*), we would need a means of transmitting a powerful narrowband signal. This is the flip side of SETI: instead of pondering how best to *listen* we consider the practicalities of how to *transmit*. Of course, by studying the problem of how to transmit a signal over interstellar distances, we can learn a lot that will help us to listen for signals.

Suppose we decide to use radio. The first problem is which transmission frequency to use. The logic that makes us *listen* for signals at the waterhole suggests we should *transmit* somewhere in that region, although arguments could be made for several other frequencies. Once we have decided on a frequency — and let us assume for the moment that we should broadcast in the waterhole — what technology would be required?

Since we do not know in advance where an ETC might reside, the safest option is to transmit isotropically — with the same power in all directions. If we wanted to send a narrowband signal so that it could be detected by a small antenna at a distance of 100 light years, say, then the power required by the transmitter would exceed the present total installed electricity-generating capacity of the world. And 100 light years barely extends beyond our immediate neighborhood. The farther away we want the signal to be received, the larger the power requirement of the transmitter. An isotropic transmitter is thus quite beyond our present technological capability. Even if we *could* build such a device, would we commit such a large level of resource to a project that has no guarantee of success?

If ETCs listen with an Arecibo-type telescope rather than a simple dish, then the power requirements for the transmitter lessen. Indeed, if we knew the precise location of an Arecibo-type telescope on the other side of the Galaxy, then our own Arecibo could send it a signal. The problem is that we do not know in advance where to point the transmitter. An Arecibo-type dish, operating at a frequency in the waterhole region, has an extremely narrow beam. The old needle–haystack dictum does not begin to convey the improbability of sending a narrow beam that just happens to align with a large receiver somewhere in the depths of space.

Isotropic transmission, which guarantees that anyone with an ear can hear you, is exceedingly expensive; beamed transmission, which is cheap, excludes most of your potential audience. These are the two extremes for a transmission strategy. We could make various trade-offs and compromises, and ETCs may be able to devote more resources to transmission than can mankind at present. But interstellar radio transmission is not easy.

In the light of these difficulties — and there are several others I have not described — maybe ETCs decide to let others do the hard work of transmission. Maybe the Galaxy is full of civilizations waiting for a call?

This is an unlikely resolution of the paradox. The difficulties may seem insurmountable to us, but they would surely present less of a challenge for, say, a K3 civilization. And many of the problems surrounding transmission are surmountable even with our *present* level of technology — if we move away from the idea of using radio waves!

Even with our present laser technology we can generate a pulse of light that, for a short duration, outshines the Sun. An advanced ETC would presumably have no trouble in generating a pulse that is, briefly, billions of times brighter than its star. Such pulses can be detected with a relatively small optical telescope connected to a charge-coupled device. Furthermore, over distances of a few thousand light years, the interstellar medium has relatively little effect on a visible light signal; unlike radio, optical communication is not corrupted. Lasers are in many ways a more effective transmission mechanism than are radio dishes.

The drawback with optical-based communication is that the beam is *extremely* narrow. The transmitting civilization must therefore know the *precise* location of the receiving telescope. It is the same problem radio transmitters face if they generate narrow-beam signals, except it is much worse. It is futile to send a laser signal at random; the beam is unlikely ever to be detected. The transmitting civilization must therefore draw up a list of target planetary systems along with precise and accurate values for the positions of those systems. Furthermore, stars are not at rest. If an ETC sends a signal to where the star is *now*, then by the time the light reaches it the star will have moved on. So the transmitting civilization also needs accurate information about the velocities of the target stars.

Gathering information about other planetary systems and the precise location and velocity of stars is not easy; but neither is it impossible for a civilization advanced beyond our own. The recent *Hipparcos* mission obtained such data on the nearest stars, and proposed projects like the ESA *Darwin* mission and NASA's *Terrestrial Planet Finder* will detect any Earth-sized planets around the nearest 200 stars.[134] If *we* can contemplate such missions, then a civilization just a little more advanced than ours should be able to use optical communication over interstellar distances — and radio signals too, if they choose. So there seems to be no technical reason why ETCs cannot transmit.

*　　　　*　　　　*

It is worth mentioning that mankind has already beamed two signals to the stars (deliberately, that is, as opposed to leakage from broadcasting stations). The first intentional signal was sent in 1974.[135] Its author was Drake, who took the opportunity to use the inaugural ceremony of the refurbished Arecibo telescope to send a message at 2.38 GHz in the direction of M13. (This is a globular cluster containing about 300,000 stars, but unfortunately not of the type we expect to possess Earth-like planets.) The message lasted 3 minutes and was only 1679 bits long, but Drake managed to pack in a lot of information. When the signal reaches M13 in about 24,000 years time, if astronomers there could decode it, they might learn a surprising amount about us. Even if they could not decode it, the very detection of the signal would convey information; it would tell them an intelligent species was here and it had advanced to the radio stage — the very fact of the signal carries a message. (The second broadcast, in 1999, was a 400,000-bit message at 5 GHz to four nearby Sun-like stars. The message was sent several times; unfortunately, the first transmission contained a typographical error.)[136]

It is also worth mentioning that Drake was criticized because he made his broadcast without consulting widely. The transmission represented Earth, yet no national governments were asked their opinion about the content of the signal. In practice, isolated transmissions like this have essentially no chance of being detected; but perhaps future large-scale transmissions from Earth will require a planetary government that can speak for us all. Perhaps an advanced ETC, recognizing the ethical problems of transmitting signals to the Universe, only transmits when it has achieved a level of unity such that its signals represent a consensus of their entire world. And perhaps that is why we are still waiting to hear from them: they listen not because of technical difficulties but because of ethical difficulties.[137]

This is another unlikely resolution to the paradox. Attributing motives to putative alien civilizations is probably futile. And once again we have to ask whether a concern over the ethical niceties of transmission would affect *every* civilization? All we can say with certainty is that sending a message to the Universe, in the expectation that it will be received by another civilization, is difficult. But it is not impossible. Some civilizations should be out there, signaling their presence. So why have we not heard from them?

SOLUTION 22:   BERSERKERS

In the long run, we are all dead.
J. M. Keynes

During the 1950s, Cold War strategists toyed with the idea of a "Doomsday weapon." Such a weapon was terrible, uncontrollable, capable of destroying all human life on Earth — including the owners of the weapon. If your enemy knew you were willing to deploy a Doomsday device then — so the Cold War logic went — they would not dare attack you. I suspect that

Chapter 4

Fred Saberhagen had the Doomsday weapon in mind when he wrote his famous berserker stories.[138]

Berserkers are sentient, self-reproducing machines that are savagely inimical to organic life. Think of them as paranoid Bracewell–von Neumann probes with a mean streak. The relevance to the Fermi paradox is clear: ETCs have either been prevented from arising by berserkers, wiped out by berserkers, or else are keeping quiet for fear of attracting berserkers. It is an elegant solution to the Fermi paradox. But could berserkers exist outside the pages of science fiction?

If an ETC could build probes capable of colonizing the Galaxy, then unfortunately berserker construction would presumably not be beyond them technically. It is hard to imagine any intelligent species actually *wanting* to develop berserkers, since the technology is so dangerous to the creators as well as to all other life. Besides, what would be their motivation for constructing berserkers? If their aim was to colonize the Galaxy for itself, then it could fulfill its aim simply by being the first to colonize: remember that the colonization time for the Galaxy is much less than the age of the Galaxy. However, we should not be overly sanguine about the prospect of berserkers. Suppose the programming of a "well-adjusted" probe mutates; perhaps a collision with a stray cosmic ray changes the line of code in its core module from "seek out new life and new civilizations" to "seek out new life and new civilizations, *and kill them*." Self-reproducing probes will inevitably evolve, and so berserker-type devices *might* develop.

*   *   *

The berserker solution has been criticized on several grounds. Even if berserkers exist, would they be an inevitable Nemesis? Could not ETCs "inoculate" themselves, much as they would inoculate themselves against a virulent disease? Most tellingly, the berserker scenario suffers from a Fermi paradox of its own: if berserkers exist, then how come *we* are here? Berserkers should already have sterilized our planet. Instead, as we shall see in later sections, the geological record indicates life has been present on Earth for billions of years. To be sure, Earth has seen several mass extinctions, but there are natural explanations for these events. (The Universe is dangerous enough without berserkers.) So why have berserkers silenced all other civilizations but left us alone? We could argue that berserkers destroy only technological life-forms and need a "trigger" — presumably the detection of radio waves — before they begin work. But that extra step in the argument spoils what is potentially an elegant resolution of the Fermi paradox. Besides, we have been using radio for a century and may soon go radio-quiet despite our burgeoning level of technology. If berserkers are all they are cracked up to be, then where are they?

SOLUTION 23: THEY HAVE NO DESIRE TO COMMUNICATE

Speech is great; but silence is greater.
Thomas Carlyle, *Essays: Characteristics of Shakespeare*

So far we have assumed that ETCs *want* to communicate. Maybe they don't?

Resolutions of the paradox based on the idea that ETCs will keep themselves to themselves depend on making assumptions about the motives of alien beings. If such beings exist, they will be the product of billions of years of evolution in unearthly environments, with senses, drives and emotions different from our own. Or they may be artificial intelligences that have taken over from their biological creators. Or they may be of a form quite beyond our imagining. Whatever the case, how can we pretend to understand the motives of intelligences so vastly different from ours? Probably we *cannot* understand alien motives — but it is fun to speculate.

*　　　　　*　　　　　*

One reason why ETCs might choose to keep quiet is fear. When an ETC broadcasts to space, it reveals its location and perhaps its level of technology. Any neighbors who are listening may be aggressive; worse still, they may be berserkers. We have no idea whether aliens would think this way, but many humans certainly do. Perhaps caution is a general trait among advanced intelligences.[139]

Others have suggested that the spirit of curiosity pervading humanity (and many other terrestrial species) might be lacking in intelligent extraterrestrials. Perhaps ETCs simply have no interest in exploring the Universe or in communicating with other civilizations. One could argue that ETCs lacking curiosity and a desire to learn how the Universe works would never develop the technology to communicate over interstellar distances; that any intelligent species we meet *must* have curiosity about the external world. But a glance through the history books shows that there have been human cultures that were isolationist, wanting nothing to do with others. Perhaps a similar philosophy is common among ETCs.

A more common argument, usually advanced in a spirit of humility, is that any ETCs would be so far beyond us intellectually they would be indifferent to our existence. I heard one astronomer say that advanced civilizations "would not want to communicate with us because we could teach them nothing; after all, we don't want to communicate with insects." Yet is that true? We are unlikely to be able to teach an advanced ETC anything about a "hard" science like physics, say. But actually, physics is easy: the Universe is constructed of a small number of building blocks that interact in a small number of well defined ways. The Universe is intelligible.

Advanced ETCs are therefore unlikely to spend much time discussing physics; they will all have the same theories of physics because they all inhabit the same Universe. The areas of study that are *really* hard — in the sense of difficult to master — are subjects like ethics, religion and art. Advanced ETCs would not want to learn about electromagnetism from us, but they might be fascinated in trying to comprehend and understand how we see the Universe — a challenge worthy of them. Furthermore, it is just not correct to say that "we don't want to communicate with insects." Biologists have gone to great lengths to interpret signals that might be encoded in the dance of the honeybee; pheromone communication by ants has long been studied. Such investigations are part of a wider study of animal communication and animal cognition. Indeed, the possibility of communication with "lower" species has fascinated humans for thousands of years. Just because we might be a "lower" species compared to others does not mean that we are inherently uninteresting. (Besides, even if ETCs *are* indifferent to lower forms like us, it does not explain why we have not seen or overheard their communications with peers.)

Not All Cultures Are Expansionist

The most frequently cited example of an isolationist civilization is that of China under the Ming dynasty.

The dynasty was founded in 1368 by Zhu Yuanzhang, who became the Hongwu emperor (which in translation means Extremely Martial).[140] Under his rule, and later that of the Yongle emperor, China expanded her empire. The Yongle emperor and his successor, the Xuande emperor, sent the great admiral and explorer Zheng He on seven remarkable voyages. The voyages took him as far as India, the Persian Gulf and the coast of East Africa. Zheng He commanded one of the greatest armadas in history — on his first voyage, 60 of the 317 ships were 400-foot "Treasure Ships"; it must have been an awe-inspiring sight — and undoubtedly China was the leading maritime power of the day. Indeed, China was probably the most technologically advanced nation on Earth. But after the deaths of Zheng He and the Xuande emperor, and for reasons that are still debated, China ceased its expansionist policies, forbade foreign trade, and embarked on an inward-looking path.

Another common argument is that super-intelligent ETCs refrain from communication with us to protect us from an inferiority complex; they are waiting until we can provide worthwhile contributions to the conversations taking place in the Galactic Club. (Presumably, therefore, the ETCs deliberately "talk above our heads"; they may also place us under interdict, as we discussed earlier.)[141] As Drake pointed out, however, on an individual basis all of us routinely deal with minds superior to our own. As children we learn from our elder siblings, parents and teachers; as adults we learn from the great authors, scientists and philosophers of the past. It's no big deal: at worst, when we find we will never write as well as Shakespeare or have insights as profound as Newton, we might be disappointed — but then we shrug, and we do the best we can. At best, viewing the accomplishments of others serves to inspire us. Why should it be different for societies?[142]

It is possible to dream up many other reasons why intelligent extraterrestrials are reserved. Maybe they reach spiritual fulfillment on their home planet and see no need to search for others. Maybe they believe only ethically advanced species should attempt to spread into space and are waiting to evolve into such a species. Maybe the inevitable time delay involved in interstellar communication makes interaction with other species appear less attractive; it would have to be one-way. (But we engage in one-way communication all the time. We read Homer because his works are *interesting*, even though we have no chance of engaging in a two-way communication with him.) Maybe — and this is a depressing thought, given our lack of progress in spaceflight since the Apollo missions — they just cannot be bothered.

*   *   *

The trouble with all such resolutions of the Fermi paradox is that they require an unlikely uniformity of motive. If the Galaxy is home to a million civilizations, as the optimists suggest, then perhaps *some* of them have no desire to communicate with others. But to explain the paradox requires *all* civilizations to behave that way. And surely that is unlikely.

Indeed, the problem might be even more acute than I suggest above. To develop interstellar communication, a civilization presumably requires a community of billions of minds. Mankind, for example, has over the centuries drawn on the genius of a vast number of minds to develop our present level of technology. If this holds true for other ETCs, then there may be trillions of intelligent individuals out there — some of whom, if they belong to a K3 civilization, will have access to unimaginably powerful technology. In this case, these resolutions of the Fermi paradox demand a uniformity of motive not only *between* ETCs but also of individual members or groups *within* an ETC!

Chapter 4

SOLUTION 24: THEY DEVELOP A DIFFERENT MATHEMATICS

The integers were created by God; all else is the work of man.
Leopold Kronecker

One of the abiding mysteries of science is, as Wigner put it, "the unreasonable effectiveness of mathematics."[143] Why should mathematics describe Nature so well? Whatever the reason, we should be grateful we can comprehend the Universe mathematically. We can build bridges that stay up, construct aircraft that remain aloft, and design computers that are a marvel of miniaturization; ultimately all modern technology is dependent upon mathematics. (People have built bridges, aircraft and computers by trial and error, but I would not want to use them.)

Perhaps most mathematicians, at least tacitly, subscribe to Platonism. The Platonic philosophy holds that mathematics and mathematical laws exist in some sort of ideal form outside the realm of space and time. The work of a pure mathematician is therefore akin to that of a gold prospector; a mathematician searches for nuggets of pre-existing absolute mathematical truth. Mathematics is discovered, not invented.

Some mathematicians, though, argue strongly from an anti-Platonic stance. They claim mathematics is not some sort of idealized essence independent of human consciousness, but is rather the invention of human minds. It is a social phenomenon, part of human culture. (This is a rather brave thing for professional mathematicians to propose, because superficially the proposal can sound like the lunatic ravings of those postmodernist critics who denounce science as the arbitrary construction of dead white male Europeans.) The anti-Platonist contends that mathematical objects are *created* by us, according to the needs of daily life. It may be, they argue, that evolution has hard-wired an "arithmetic module" into our brains. Neuroscientists even have a possible location for this module: the inferior parietal cortex, an as yet poorly understood area of the brain.[144]

I would not be surprised if we all have an arithmetic processing unit in our heads. After all, our ancestors lived in a world of discrete objects in which the ability to recognize numbers of predators or numbers of prey would have been extremely advantageous. In fact, since the ability to make rapid judgments based upon the perceived numbers of objects is so clearly useful, we might expect animals to possess some sort of "number sense." And, indeed, there is evidence that rats and raccoons, chickens and chimpanzees can make rudimentary numerical judgments. (However, it is unlikely that any animal other than humans can *count* in the sense that we understand it. In experiments claiming to demonstrate counting ability in animals, it is difficult to rule out the possibility that animals are using much simpler cognitive processes. For example, when small numbers of objects are involved, the animals might be subitizing. We do the same ourselves: if we are presented with a plate containing 3 biscuits we *know* there are 3 biscuits, not 2 or 4, without having to count them. This is subitizing — a perceptual process that works for numbers of objects up to about 6. The process works well for 3 objects, say, because there is only a limited number of ways of arranging them [variations on the patterns ∴ and ⋯ pretty much exhaust the possibilities]. There are so many ways of arranging 23 objects, say, that no perceptual clue enables us to readily distinguish a group of 23 objects from 22 or 24 objects. Similarly, many animals can judge relative numerousness. They will prefer a large quantity of food to a smaller quantity, for example. Again, though, the animals need not count — after all, a pile of 500 birdseeds simply looks bigger than a pile of 300 birdseeds. In such experiments animals are almost certainly using visual cues to distinguish between situations.) So, although the ability to do integral calculus, or even simple multiplication, is not innate, one might argue that the foundations of arithmetic — from which the worldwide community of mathematicians has constructed such a wonderful edifice of abstract thought — *are* innate. The integers are not ideal Platonic forms existing independently of human consciousness; rather they are creations of our minds, artifacts of the way the brains of our ancestors interpreted the world around them.

If this is correct, then a fascinating question arises: what would the mathematics of an ETC be like? (Their symbols, of course, would be different; but that is not important.) Would they have developed the prime number theorem; the min–max theorem; the four-color theorem? If their evolutionary history were completely different from our own, then perhaps not. Why should it?[145] If they evolved in an environment in which variables changed continuously rather than discretely, then perhaps they would not invent the concept of an integer. Or perhaps it is possible to develop a mathematical system based upon the concepts of shape and size, rather than number and set as humans have done. I personally find it difficult to imagine such alien mathematics, but that is almost certainly a deficiency in my imagination; it does not prove that it is impossible for such different systems to exist.[146]

None of this is to say that our own mathematics is *wrong*. Surely the relation $e^{\pi i} = -1$ is true and unavoidable anywhere in the Universe. (At least, I do not see how it could be otherwise.) But other intelligences, which have a different evolutionary history, may simply not see the relevance of concepts like e or π or i or -1. Equally, they may have concepts — important in their own environments — that we have failed to invent.

The point here is that human mathematics enabled us to develop technology. Perhaps this type of mathematics is *required* for the development of technology. For a civilization to build radio transmitters capable of broadcasting over interstellar distances, it simply has to understand the inverse-square law and a host of other "terrestrial" mathematics. A solution to the Fermi paradox, then, could be that other civilizations develop other systems of mathematics — systems that are inapplicable for use in building interstellar communication or propulsion devices.

*   *   *

As a resolution to the paradox this suffers from the same difficulty as several others: even if it applies to *some* civilizations (and many would deny even that possibility), it surely cannot apply to *all* civilizations. I can imagine a race of intelligent ocean-dwelling creatures developing a mathematical system without the Pythagorean theorem (would they even know about right angles?), but not *every* species will live in the ocean; some will be land creatures, like us, and it seems reasonable to suppose that at least some of them would develop familiar mathematics.

One final thought. Mathematics, at its heart, is about patterns. Even if mathematics itself is universal, perhaps different intelligences appreciate and investigate different types of pattern. There could be nothing more interesting for mathematicians than to learn about different mathematical systems. To me, this makes it seem even more probable that ETCs would want to communicate with each other.

SOLUTION 25: THEY ARE CALLING BUT WE DO NOT RECOGNIZE THE SIGNAL

I really do not see the signal.
Nelson, at the Battle of Copenhagen

There is a more subtle argument relating to the previous section. Suppose advanced ETCs do indeed create "different" mathematics, or — which is easier to accept and may amount to the same thing — suppose their mathematics was millions of years in advance of ours. If they were transmitting to us right now, would we even recognize that their transmissions were artificial?

Much of the present SETI effort concentrates on the waterhole region and on simple multiples of the hydrogen line frequency (2, 3, π times the frequency, and so on). Perhaps ETCs using a different mathematics see nothing special about such frequencies; the "obvious" frequencies for them might be something quite different. But that is a minor point; let us suppose they broadcast in the waterhole region. The hope of communicating with

ETCs is predicated on finding signals containing simple mathematical patterns and developing from this a shared language. In other words, we hope to receive signals encoded in some math-based language like Freudenthal's LINCOS.[147] Is this hope reasonable?

There are two aspects to this question, which we should keep separate. First, could we recognize a signal as artificial? Second, if we recognize a signal, could we decode its meaning?

The efforts of SETI scientists are doomed if they cannot distinguish between an artificial transmission and a natural emission. However, physicists have shown that if a message is sent electromagnetically and has been encoded for optimal efficiency, then an observer who is ignorant of the coding scheme will find the message indistinguishable from black-body radiation.[148] Now, black-body radiation is simply the radiation an object emits because it is hot; astronomers detect black-body radiation all the time, and of course they apply the simplest explanation to their observations. But they *could* be observing messages that have been encoded for optimal efficiency! If ETCs do not care whether we know about them, and if they encode their communications to each other with optimal efficiency, then we could intercept their messages and remain unaware of their existence. It is yet another difficulty that SETI scientists must face.

If advanced ETCs *want* us to find them, then they could easily encode messages we would recognize as artificial. A signal containing pulses distributed according to some obvious pattern — the first few prime numbers, say — would leave no doubt in our minds about its origin. We have to hope, then, that ETCs *want* to be noticed. But even if we detect a message, could we decode the contents? Consider the Voynich Manuscript.[149] In 1912, Wilfred Voynich, a collector, bought this 234-page book from the Jesuit College at the Villa Mondragone, Frascati, in Italy. It presently resides in the Rare Book Room and Library of Yale University, where it is cataloged by the less romantic name of MS 408. The book was probably written some time between the 13th century and 1608. And this is almost everything we know about the manuscript: it was written in a language or code that no one has yet deciphered. It seems to contain information about herbalism and astrology, among other things, but no one is sure; it could, for example, be a medieval hoax.

Whatever information the Voynich Manuscript contains, we know it was written by a human being in the not too distant past. So the author had the same sensory inputs as the rest of us; a cultural background that is recognizable, if not identical to our own; human emotions that drove him in exactly the same way that they drive us. And yet he wrote a book we cannot decipher. If such a situation can occur with a member of our own species, what chance do we have of understanding a message from an ETC?

FIGURE 38   Folio 78r from the Voynich Manuscript. Note the strange text characters. At first glance they seem to be from a foreign language that you cannot quite place; but detailed researches have shown that the characters belong to no known language. Are they characters in some private code? Is the whole thing simply a hoax? No one is sure.

If aliens exist, then they will possess different sense organs, different emotions, different philosophies and, perhaps, even different mathematics. I suspect if astronomers ever detect a message from intelligent extraterrestrials, the dominant emotion mankind would feel — after an initial period of excitement and euphoria — would be frustration. We might struggle for millennia without ever deciphering the meaning of the message.

But does any of this have relevance to the Fermi paradox? Well, one scenario people have offered is that ETCs long ago realized interstellar travel was impossible, made contact with each other through EM signals and, over the aeons, agreed to communicate with each other with signals encoded for optimal efficiency. They then lost interest in contacting younger civilizations like our own, so we find the Galaxy filled with black-body radiation. That may have happened, but it is another example of a "just-so" story; it offers no testable prediction.

On the other hand, if we detected a signal that was clearly artificial in origin, then, even if we could not decipher it, we could infer the existence of intelligent extraterrestrial beings. So whether we can understand aliens is a quite separate question from whether they exist and has no real bearing on the Fermi paradox.

SOLUTION 26: THEY ARE SOMEWHERE BUT THE UNIVERSE IS STRANGER THAN WE IMAGINE

> listen: there's a hell
> of a good universe next door; let's go.
> e. e. cummings, *pity this busy monster, manunkind*

The theories of modern physics are remarkable in their range of applicability. They explain phenomena at scales as small as the electron and as large as super-clusters of galaxies. They explain events that happened a tiny fraction of a second after the Big Bang, and we can use them to determine the fate of the Universe.

Some might say physicists are arrogant, filled with hubris for daring to claim such success for their theories; science, being the product of the human brain, cannot possibly capture the subtleties and mysteries of the Universe. In my experience such people tend to accept the UFO explanation of the Fermi paradox. However, a few scientists and many SF writers have offered some interesting suggestions. They explain the paradox by supposing that the Universe is not quite what physicists think it is.[150]

For example, perhaps intelligent species evolve to a non-physical state that transcends the limitations of spacetime. Clarke's novel *Childhood's End* describes mankind's transition from our present immature state to a merger with the Galactic "overmind" (some sort of spiritual union, the precise nature of which is not made clear). According to this suggestion, we do not hear from ETCs because they have evolved beyond our secular existence.

Another suggestion: all other intelligences evolve telepathic abilities and can communicate directly, from mind to mind, even over interstellar distances. Not for them the difficulties of radio communication. Perhaps they even *travel* using the power of the mind — like the jaunt in Bester's novel *The Stars My Destination*. If this were true, ETCs might be unaware of our psi-challenged existence.

Yet another suggestion, just as outrageous but based on more conventional ideas, is that ETCs are busy exploring parallel universes. The *many-worlds interpretation* of quantum mechanics suggests that every time we make a measurement on a quantum system possessing two possible states, the Universe splits — into universe A and universe B.[151] An observer in universe A measures one outcome of an experiment, an observer in universe B measures the other possible outcome. The result is a never-ending branching of universes. In the totality of universes, all possibilities are realized. If the many-worlds interpretation is correct (a big "if" — there are several competing interpretations of quantum mechanics, and there is no direct evidence in favor of the many-worlds interpretation) and if it is possible to move between universes (an absolutely huge "if" — there is no indication that such travel could occur), then perhaps ETCs are elsewhere. Why stick around a dull place like this Universe when you can explore *really* interesting places?

*   *   *

While it is certainly true that science has not told us *everything* — indeed, what remains to be discovered seems to grow exponentially — it is wrong to say science has told us *nothing*. The Universe seems to be intelligible; and over the past 400 years our science — a process involving hundreds of thousands of people working individually and cooperatively — has yielded reliable knowledge about the Universe. Any new theories not only have to explain new observations and experimental findings, but also the accumulated set of observations and findings — which makes it extremely tough to develop new theories. No one has succeeded in developing useful theories of phenomena like transcendent spiritual unions, interstellar telepathic communication, inter-universe travel — or any of the other imaginative suggestions that SF writers have made. In fact, since at present we can understand the Universe without invoking the existence of such phenomena, we do not *need* to develop new theories to explain them. (That does not mean such phenomena are impossible; but we require evidence before we need to study them in earnest.)

So while these suggestions all make for good stories, it is difficult to take them seriously as resolutions of the Fermi paradox.

SOLUTION 27: A CHOICE OF CATASTROPHES

> *. . . we make guilty of our disasters the sun, the moon, and the stars; as if we were villains on necessity, fools by heavenly compulsion.*
> William Shakespeare, *King Lear*, Act I, Scene 2

An obvious, if gloomy, resolution of the Fermi paradox occurs if L — the factor denoting the lifetime of the communicating phase of an ETC — is small. In Chapter 5 I shall deal with various ways in which Nature is hostile to life. Here, though, I want to examine the idea that intelligent species may be the inevitable authors of their own doom.[152]

War

FIGURE 39   *A 350-kTon thermonuclear explosion (mid-1950s).*

To more than a few scientists working during the Cold War it seemed quite certain that ETCs would discover the interesting properties of element 92 (known to us as uranium) and therefore learn how to construct nuclear weapons. For several scientists, then, the reason for a short lifetime (in other words, a small value for L) was obvious: all advanced civilizations inevitably annihilate themselves in a nuclear holocaust, as the human race was apparently on the verge of demonstrating.[153]

It hardly seems worth mentioning that, depending upon the severity of a nuclear war, the extinction of an intelligent species might follow.[154] The world's arsenals still contain many thousands of nuclear weapons, and if they were ever used in large numbers, then they would certainly destroy *Homo sapiens*. A limited nuclear war might be just as ruinous for our species, due to the effects of a potential global nuclear winter.[155]

Nevertheless, as many SF writers have demonstrated, it is possible to imagine scenarios in which members of a warring species survive a limited war and, over a period of thousands of years, recreate their civilization. One of the earliest post-apocalyptic novels, and certainly one of the best, is Miller's *A Canticle for Liebowitz*. Miller describes how a flicker of knowledge is preserved by monks after a nuclear war has decimated the population.[156] In *Canticle*, mankind eventually rediscovers the power of science and, a few millennia after the first nuclear holocaust, has "advanced" to the stage where the Bomb can be dropped once again. Is the urge to war so deeply ingrained that a civilization learns nothing? Are civilizations somehow *compelled* to drop bombs as soon as they can? Unless that is the case, limited nuclear war cannot provide an explanation of the paradox. It may take many thousands of years to recover a high level of civilization after limited nuclear war, but this timescale is short — just a few minutes of the Universal Year.

Even a total, all-out, no-holds-barred nuclear war would not destroy *all* life on a planet. Consider the organism *Deinococcus radiodurans*. Scientists first isolated it in 1956 from a can of ground beef; the beef had been radiation-sterilized, but the meat still spoiled. It turns out that *D. radiodurans* can survive an exposure to gamma-radiation of 1.5 million rads. (For comparison, a dose of 1000 rads is usually enough to kill a man.) Exposure to intense radiation blasts apart its DNA — but within a few hours the organism reforms its entire genome with seemingly no deleterious effects. This organism can withstand other extreme conditions, such as prolonged desiccation, which explains why it is often called "Conan the Bacterium." A nuclear war would not unduly inconvenience Conan the Bacterium. And not just bacteria would survive; various other organisms would live through a nuclear war. If intelligence is an inevitable outcome of evolution (this is contentious, as we shall see, but it is presumably the viewpoint of those who argue there are a million ETCs in the Galaxy,) then the wait for intelligence to emerge after a nuclear holocaust would not be endless: a few hundreds of millions of years, perhaps. This is an unimaginably vast reach of time on a human scale, but, again, it is not particularly significant when compared to the age of the Galaxy — a few days in the Universal Year.

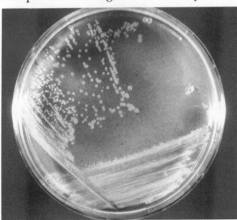

FIGURE 40 The organism Deinococcus radiodurans *growing on a nutrient agar plate. This bacterium can survive extremes of radiation and desiccation.*

Those civilizations that avoid the Scylla of nuclear war must still navigate the Charybdis of biological and chemical warfare. For example, we know that chemical weapons can be used to destabilize ecosystems; genetically engineered biological weapons can destroy food supplies or decimate populations directly. But the comments made above regarding nuclear war also hold for these forms of warfare. Is it likely that *every* ETC civilization, when it reaches a certain stage (and before it establishes colonies in space), annihilates itself through warfare? Without wishing to tempt fate, we can hope that *Homo sapiens* has shown that at least one species in the Galaxy can resist the urge to self-destruct through war.

Overpopulation

One of the defining characteristics of life on Earth is reproduction. Presumably this is a universal characteristic of life. If we ever meet the equivalent of the Krell from *Forbidden Planet*, the Soft Ones from *The Gods Themselves*, or the Greeshka from *A Song for Lya*, then we may be surprised by the mechanics of their reproductive processes — but not the fact that they reproduce. And since aliens will reproduce, they will be subject to the same simple mathematical laws that describe population growth here on Earth.

Until about 8000 BC, the number of people on Earth at any time never exceeded about 10 million people. Health was poor and living conditions were harsh; life expectancy at birth was probably 30 years or less. Had the birth rate not been as high as the death rate, human society would have died out; for the continued existence of families, clans and tribes it was vital that adults had as many children as possible. Even so, the rate of population growth was barely above zero. The situation began to change when mankind developed agriculture. Life expectancy began to increase under an agricultural lifestyle, and the birth rate began to exceed the death rate. (People are generally very quick to adopt new technologies; social attitudes, such as "be fruitful and multiply," are slower to change. So although the reasons for having large families had lessened, social pressures on parents had not.) Fortunately, agriculture supported a greater population than the old hunter–gatherer way of life; by 1650, the world's population was 0.5 billion — a 50-fold increase over the steady-state population size for 99% of human history. By around 1800, the world's population reached its first billion — a doubling in 150 years. By 1930, the population was 2 billion — a doubling in 130 years. By 1975, the population was 4 billion — a doubling in just 45 years. The world's population exceeded 6 billion in September 1999.

To say that this past rate of population growth cannot continue is to risk being labeled a Cassandra. But it *cannot* continue. *Really.* At those growth rates, in a few hundred years the combined flesh of humanity would form a sphere expanding at the speed of light. (Of course, this would not happen; if we did not slow the growth rate, then biology would curb it for us long before relativistic effects become apparent.)

If we are lucky, the world's population will in the next few decades reach a new steady state, with a low death rate matched by a low birth rate. (Though even this would not satisfy the Cassandras, since there are downsides to this situation. For example, the elderly would consume a large share of costly public services, while there would be fewer young people to work and pay for them.) The steady-state population will probably be in the range 11 to 13 billion. Whether Earth can feed so many people and offer them a reasonable standard of life is not known. But even if it can, what damage will 13 billion people inflict upon it? A much smaller population has managed to transform or degrade up to half of Earth's land surface for agricultural and urban use; we have increased the atmospheric CO_2 concentration at an alarming rate; we already use more than half of the accessible surface's fresh water; the natural rate at which species become extinct has accelerated due to human activity; and so on, and so on. None of these problems (not to mention problems such as poverty and social injustice) are caused solely by overpopulation; but overpopulation certainly does not help in the search for solutions.

Since alien life will reproduce, it seems inevitable that at some stage an ETC will face a population crisis. But will *every* civilization fail to negotiate the crisis?

The Gray Goo Problem

Nanotechnology seems as if it might be the natural outcome of converging advances in many different fields of knowledge.[157] The term refers to engineering that takes place at the nanoscale, a scale where the dimensions of objects are typically measured in nanometers (billionths of a meter). Since molecules are of this size, it also goes by the name of molecular engineering. Future nanotechnologists will have the ability to assemble custom-made molecules into large, complex systems; their capacity to create materials will be almost magical. (Since this capacity appears to be so wonderful, and yet is presently far beyond our abilities, several commentators are skeptical of nanotechnology. So it is worth emphasizing that there seems to be no fundamental reason why we cannot develop the technology. Nature herself is a "nangeneer": enzymes, for example, are nanotechnological devices that employ biochemical techniques to carry out their tasks. If Nature can do it, so can we. It is also worth emphasizing that the success or failure of nanotechnology will determine whether we ever develop Bracewell–von Neumann probes.)

One of the elements of any future nanotechnology is likely to be the *nanorobot* — or nanobot, for short. Although their development is a long way off, theoretical studies suggest we could construct nanobots from one of several materials — with carbon-rich diamondoid materials perhaps forming the basis for many types of nanobot. Studies also suggest that one of the most useful types of nanobot will be a self-replicating machine.

Alarm bells start to ring whenever self-replication is mentioned. The danger inherent in producing a self-replicating nanobot in the laboratory is clear upon answering the following question: What happens when a nanobot escapes into the outside world? In order to replicate, a nanobot made of a carbon-rich diamondoid material would need a source of carbon. And the best source of carbon would be the Earth's surface biosphere: plants, animals, humans — living things in general. The swarms of nanobots (for soon there would be many copies of the original) would dismantle the molecules in living material and use the carbon to produce more copies of themselves. The surface biosphere would be converted from the rich, varied environment we see today into a sea of ravenous nanobots plus waste sludge. This is the gray goo problem.

As mentioned above in the discussion on overpopulation, exponential growth is a powerful thing. Freitas has shown that, under ideal conditions, a population of nanobots growing exponentially could convert the entire surface biosphere in less than three hours![158] We can add this, then, to the depressing list of ways in which the lifetime of the communicating phase of an ETC might be shortened: a laboratory accident, involving the escape of a nanobot, turns their biosphere into sludge.

This solution to the paradox, which has been seriously proposed, suffers the same problem as many other solutions: even if it can occur it is not convincing as a "universal" solution. Not every ETC will succumb to the gray goo.

*   *   *

The young boy in Woody Allen's *Annie Hall* becomes depressed at the thought that the Universe is going to die, since that will be the end of everything. I am becoming depressed writing this section, so to cheer up myself — and any young Woodys that might be reading — I think we have to ask whether the gray goo problem is even remotely likely to arise. As Asimov was fond of pointing out, when man invented the sword he also invented the hand guard so that one's fingers did not slither down the blade when one thrust at an opponent. The engineers who develop nanotechnology are certain to develop sophisticated safeguards. Even if self-replicating nanobots were to escape or if they were released for malicious reasons, then steps could be taken to destroy them before catastrophe resulted. A population of nanobots increasing its mass exponentially at the expense of the biosphere would immediately be detected by the waste heat it generated. Defense measures could be deployed at once. A more realistic scenario, in which a population of nanobots increased its mass slowly, so the waste heat they generated was not immediately detectable, would take years to convert Earth's biomass into nanomass. That would provide plenty of time to mount an effective defense. The gray goo problem might not be such a difficult problem to overcome: it is simply one more risk that an advanced technological species will have to live with.

127

Chapter 4

Particle Physics — A Dangerous Discipline?

In 1999, the London *Times* reported that experiments at the new Relativistic Heavy Ion Collider (RHIC) on Long Island might trigger a catastrophe. Physicists at the RHIC accelerate gold nuclei to high energies and then smash them into each other; it is an effective way of learning about the fundamental constituents of matter. The RHIC experiments, it was suggested, might destroy Earth. This immediately led some to suggest another of the "doomsday" solutions to the Fermi paradox: advanced civilizations learn to experiment in high-energy particle physics, and destroy themselves when an experiment goes wrong.[159]

Such concerns are not new. In 1942, Teller wondered whether the high temperatures in a nuclear explosion might trigger a self-sustaining fire in Earth's atmosphere. Calculations by physicists, including Fermi, put minds to rest: a nuclear fireball cools too quickly to set the atmosphere on fire.

The flurry of concern with the RHIC began when someone calculated

FIGURE 41 *Physicists study particle interactions at laboratories like CERN. Particles are accelerated to high energies in circular tunnels deep underground, and are then smashed into each other. (The CERN tunnels, like the one shown here, are underneath the Jura mountains.) Neither at CERN nor at RHIC are the energies remotely high enough to pose a threat to our existence.*

that the energies involved in the experiments would be enough to create a tiny black hole. The fear was that the black hole would tunnel down from Long Island to Earth's center and proceed to devour our planet. Fortunately, as more sensible calculations quickly showed, there is essentially no chance of this happening. To create the smallest black hole that can exist requires energies about 10 million billion times greater than the RHIC can generate.[160] (Even if a particle accelerator *could* generate such energies, the black hole it produced would be a puny thing indeed, with only a fleeting existence. It would struggle to consume a proton, let alone Earth.)

So we can sleep soundly, safe in the knowledge that the RHIC will not produce a black hole. We can rest assured, too, that it will not destroy Earth through the production of *strangelets* — chunks of matter containing so-called strange quarks in addition to the usual arrangement of quarks.[161] So far no one has seen strangelets, but physicists wondered whether experiments at the RHIC might produce them. If strangelets were produced, then there is a risk they might react with nuclei of ordinary matter and convert them into strange matter — a chain reaction could then transmute the entire planet into strange matter. However, having raised the possibility of catastrophe, physicists were quick to reassure everyone. Calculations show that strangelets are almost certainly unstable; even if they are stable, the RHIC would almost certainly not have the energy to create them; and even if they were created at the RHIC, their positive charge would cause them to be screened from interactions by a surrounding electron cloud.[162]

The unlikely litany of catastrophes that the RHIC (and other particle accelerators) might inflict upon us does not end with black holes and strangelets. Paul Dixon, a psychologist with only a hazy grasp of physics, believes collisions at the Tevatron particle accelerator at Fermilab might trigger the collapse of the quantum vacuum state.

A vacuum is simply a state of least energy. According to current cosmological theories, the early Universe may have briefly become trapped in a metastable state: a false vacuum. The Universe eventually underwent a phase transition into the present "true" vacuum, unleashing in the process a colossal amount of energy — it is similar to what happens when steam undergoes a phase transition to form liquid water. But what if our *present* vacuum is not the "true" vacuum? Rees and Hut published a paper in 1983 suggesting this could be the case.[163] If a more stable vacuum exists, then it is possible for a "jolt" to cause our Universe to tunnel to the new vacuum — and the point at which the jolt occurs would see a destructive wave of energy spread outward at the speed of light. The very laws of physics would change in the wake of the wave of true vacuum.

Dixon thought that experiments at the Tevatron might cause a jolt that could collapse the vacuum. He was so worried he took to picketing Fermilab with a homemade banner saying "Home of the next supernova."[164] Once again, however, we need not worry unduly about an accelerator-induced apocalypse. As Rees and Hut themselves pointed out in their original paper, through the phenomenon of cosmic rays Nature has been carrying out particle-physics experiments for billions of years at energies much higher than anything mankind can achieve.[165] If high-energy collisions made it possible for the Universe to tunnel to the "true" vacuum — well, cosmic rays would have caused the tunneling to occur long ago.

The concept of an accelerator accident causing the destruction of a world (or the whole Universe, in the case of a vacuum collapse) is really a non-starter. The physics of these events is not known perfectly — that is why physicists are carrying out the research — but they are well enough known for us to realize that the doom-merchants have it wrong in this case. We have to look elsewhere for a resolution of the paradox.

Doomsday and the Delta t Argument

There are many ways in which mankind might destroy itself. In addition to the calamities discussed above, one could add genetic deterioration, over-stabilization, epidemics or a dozen other problems. And this is without mentioning the many external factors that threaten us, such as meteor impact, solar variability and gamma-ray bursters. It barely seems worth getting out of bed in the morning. Surely, though, an intelligent species like *Homo sapiens* will learn how to navigate these problems? Remarkably, there is a line of reasoning, called the delta t argument, that suggests not.

*　　　　　*　　　　　*

In 1969, when he was a student, Richard Gott visited the Berlin Wall. He was on vacation in Europe at the time, and his visit to the Wall was one of several excursions; he had seen the 4000-year-old Stonehenge, for example, and was suitably impressed. As he looked at the Wall, he wondered whether this product of the Cold War would stand as long as Stonehenge. A politician skilled in the nuances of Cold War diplomacy and knowledgeable about the relative economic and military strength of the opposing sides might have made an informed estimate (which, judging by the track record of politicians, would have been wrong). Gott had no such special expertise, but he reasoned in the following way:[166]

First, he was there at a random moment of the Wall's existence. He was not there to see the construction of the Wall (which happened in 1961), nor was he there to see the demolition of the Wall (which we now know happened in 1989); he was simply there on vacation. Therefore, he continued, there was a 50:50 chance that he was looking at the Wall during the middle two quarters of its lifespan. If he was there at the *beginning* of this interval, then the Wall must have existed for $\frac{1}{4}$ of its lifespan, and $\frac{3}{4}$ of its lifespan remained. In other words, the Wall would last 3 times as long as it already had existed. If he was there at the *end* of this interval, then the Wall must have existed for $\frac{3}{4}$ of its lifespan, and only $\frac{1}{4}$ was left. In other words, the Wall would last only $\frac{1}{3}$ as long as it already had existed. The Wall was 8 years old when Gott saw it. He therefore predicted, in the summer of 1969, that there was a 50% chance of the Wall lasting a further $2\frac{2}{3}$ to 24 years ($8 \times \frac{1}{3}$ years to 8×3 years). As anyone who saw the dramatic television pictures will remember, the Wall came down 20 years after his visit — within the range of his prediction.

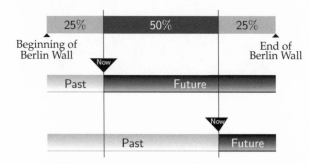

FIGURE 42　An illustration of Gott's prediction that the Berlin Wall would last for another 2 years 8 months to 24 years after he first saw it in 1969.

Gott says the argument he used to estimate the lifetime of the Berlin Wall can be applied to almost anything. If there is nothing special about your observation of a thing, then, in the absence of relevant knowledge, that thing has a 50% chance of lasting between $\frac{1}{3}$ to 3 times its present age.

In physics, it is standard practice to talk about predictions that have a 95% chance of being correct, rather than a 50% chance. Gott's argument remains the same, but there is a slight change in the numbers: if there is nothing special about your observation of an entity, then that entity has a 95% chance of lasting between $\frac{1}{39}$ to 39 times its present age. (It is important when applying Gott's rule to remember that the observation must not have any particular significance. Imagine you have been invited to a wedding and, at the reception, you start chatting to a couple you have never seen before. If they tell you they have been happily married for ten months, then you can inform them their marriage has a 95% chance of lasting between just over a week to $32\frac{1}{2}$ years. On the other hand, you can predict nothing about how long bride and groom will be together: you are at the wedding precisely in order to observe the beginning of the marriage. The flaw in applying the rule to funerals should be obvious.)

FIGURE 43   A hole in the Wall. There is a remarkable argument that links the lifespan of the Berlin Wall to the lifespan of our species!

Using the delta *t* argument to estimate the longevity of concrete walls and human relationships is amusing, but we can use it to estimate something more serious: the future longevity of *Homo sapiens*. Recent research suggests our species is about 175,000 years old. Applying Gott's rule, we find there is a 95% chance that the future lifetime of our species is between about 4500 years and 6.8 million years. That would make the longevity of our species somewhere between about 0.18 and 7 million years. (Compare this with the average longevity for mammalian species, which is about 2 million years. Our closest relatives, *Homo neanderthalensis*, survived for maybe 200,000 years; *Homo erectus*, another Hominid species and possibly one of our direct ancestors, lasted for 1.4 million years. So Gott's estimate is certainly in the right ballpark for species longevity.) The argument does not say *how* we are going to meet our end; it could be by one or more of the methods discussed above, or by something quite different. The argument simply says that it is highly likely our species will perish some time between 4500 years and 6.8 million years from now.

If this is the first time you have met Gott's argument, then you may well think (as I confess I did) that it is nonsense. However, it is difficult to pinpoint exactly where the logic is faulty. The "obvious" objections to the argument have been robustly refuted. Before examining possible objections to Gott's line of reasoning, and looking at the implications of the delta t argument for the Fermi paradox, it is worth considering a slightly different version of the same idea.

Imagine you are a contestant on a new TV game show. The rules of the game are simple. Two identical urns are put in front of you and the host tells you one urn contains 10 balls and the other contains 10 million balls. (The balls are small.) The balls in each urn are numbered sequentially (1, 2, 3, . . . , 10 in one urn; 1, 2, 3, . . . , 10,000,000 in the other). You take a ball at random from the right urn and find the ball is number 7, say. The point of the game is for you to bet whether the right urn contains 10 balls or 10 million. The odds are not 50:50. Clearly, it is far more likely that a single-digit ball comes from the urn with 10 balls than from the urn with 10 million. Surely, you would bet accordingly.

Now, instead of two urns consider two possible sets of the human race, and instead of numbered balls consider individual human beings numbered according to their date of birth (so Adam is 1, Eve is 2, Cain is 3, and so on). If one of these sets corresponds to the real human race, then my personal number will be about 70 billion — as will be any of the readers of this book, since of the order of 70 billion people have lived since the beginning of our species. Now use the same argument as we did with the urns: it is much more likely you will have a rank of 70 billion if the total number of humans who will ever live is, say, 100 billion than it is if the total number is 100 trillion. If you were forced to bet, you would have to say it is likely only a few more tens of billions of people will live. (A few tens of billions of people sounds a lot, but at the present rate we add a billion people to Earth's population every decade.)

The delta t argument is an extension of the Copernican principle. The traditional Copernican principle says we are not located at a special point in space; Gott argues we are not located at a special point in time. An intelligent observer, such as you, Gentle Reader, should consider yourself to be picked at random from the set of all intelligent observers (past, present and future), any one of whom you could have been. If you believe mankind will survive into the indefinite future, colonize the Galaxy, and produce 100 trillion human beings, you have to ask yourself: why is it that I am lucky enough to be among the first 0.07% of people who will ever live?

Gott uses the same type of probabilistic argument to deduce a variety of features of Galactic intelligence, some of which are directly relevant to the Fermi paradox. They all depend upon the idea that you are a random intelligent observer — with no special location in either space or time. First, the colonization of the Galaxy cannot have occurred on a large scale by ETCs (because if it had, you — yes, *you* — would probably be a member of one of those civilizations). Second, applying the delta t argument to the past longevity of radio technology on Earth and combining this with the Drake equation, Gott finds at the 95% confidence level that the number of radio-transmitting civilizations is less than 121 — and possibly much less than this, depending upon the parameters fed into the Drake equation. Third, if there is a large spread in the populations of ETCs, then you probably come from an ETC having a population larger than the median. Thus, ETCs with populations much larger than our own must be rare — rare enough that their individuals do not dominate the total number of beings, otherwise you would be one of them. From which we deduce there is probably not a K2 civilization to be found in the Galaxy, nor a K3 civilization anywhere in the observable Universe.

As I indicated earlier, there seems to be something not quite right with the argument; it *feels* wrong — but where exactly is it wrong? There are philosophical opinions both for and against Gott's doomsday argument, and perhaps the safest course of action is to let the philosophers slug it out. Personally, though, I am uneasy with the assumption that intelligent species necessarily have a finite lifespan; recent observations indicate the Universe may expand forever, in which case it is *possible* for mankind to survive forever (in which case a straightforward application of a doomsday argument is problematic). What is the definition of mankind in this case anyway? When, exactly, does Gott believe mankind "started"? And if our species evolves into something else, does that count as the *end* of mankind?

*   *   *

This section has discussed one of the most frequently proffered solutions to the Fermi paradox: ETCs do not long stay in the radio-transmitting phase — much less the colonization phase — because they perish. There is a variety of ways this can happen, but are any of them *inevitable*? For this explanation to work, catastrophe must be unavoidable.

SOLUTION 28: THEY HIT THE SINGULARITY

> *Things do not change; we change.*
> Henry David Thoreau, *Walden*

Back in 1965, Gordon Moore — the co-founder of the Intel Corporation — remarked how the number of transistors per square inch that one could fit on an integrated circuit seemed to double every 18 months.[167] This remark became known as Moore's law, though of course it is more an observation than a law of Nature. In its current incarnation, Moore's law states that data density doubles every 18 months. The law has held true in the 36 years since its formulation, and certain other computing hardware performance measures have kept pace. The result: cheap, fast computing power is readily available — and it has changed our world. If the law continues to hold over the next decade, and there seems to be no reason why it should not, then we will continue to see even faster and more powerful machines.[168]

Vernor Vinge, extrapolating the improvements in computer hardware and other technologies over the next few decades, argues mankind will likely produce super-human intelligence some time before 2030.[169] He considers four slightly different ways in which science might achieve this breakthrough. We might develop powerful computers that "wake up"; computer networks, like the Internet, might "wake up"; human–computer interfaces might develop so users become super-humanly intelligent; and biologists may develop ways of improving the human intellect. Such a super-intelligent entity might be mankind's last invention, because the entity itself could design even better and more intelligent offspring. The doubling time of 18 months in Moore's law would steadily decrease, causing an "intelligence explosion." A quicker-than-exponential runaway event might end the human era in a matter of a few hours. Vinge calls such an event the *Singularity*.[170]

The term Singularity is unfortunate, in that mathematicians and physicists already use it in a specific sense: a singularity occurs when some quantity becomes infinite. At Vinge's Singularity, however, no quantity need become infinite. The name nevertheless captures the essence of what would be a critical point in history: things would change *very* rapidly at the Singularity, and — like the singularity in a black hole — it becomes hard to predict what happens after hitting it. The super-intelligent computers (or the super-intelligent humans or human–computer beings) turn into ... what? It is difficult, perhaps impossible, to imagine the capabilities and motives and desires of entities that are the product of this transcendental event.[171]

Vinge argues that *if* the Singularity is possible, then it *will* happen. It has something of the character of a universal law: it will occur whenever intelligent computers learn how to produce even more intelligent computers. If ETCs develop computers — since we routinely assume they will develop radio telescopes, we should assume they will develop computers — then the Singularity will happen to them, too. This, then, is Vinge's explanation of the Fermi paradox: alien civilizations hit the Singularity and become super-intelligent, transcendent, unknowable beings.

Vinge's speculations about the Singularity are fascinating. And as an explanation of the Fermi paradox, the suggestion improves on explanations requiring a uniformity of motive or circumstance. Not *every* ETC will blow itself up, or choose not to engage in spaceflight, or whatever. But we can argue reasonably that every technological civilization will develop computing; and if computing inevitably leads to a Singularity, then presumably all ETCs will inevitably vanish in a Singularity. The ETCs are there, but in a form fundamentally incomprehensible to non-super-intelligent mortals like us. Nevertheless, as an explanation of the paradox, I think it has problems.

First, even if high intelligence *can* exist on a non-biological substrate, the Singularity might never happen.[172] There are several reasons — economic, political, social — why a Singularity might be averted. There are also technological reasons why the Singularity might not occur. For example, for the attainment of the Singularity, advances in software will be at least as important as hardware advances. Without *much* more sophisticated software than we currently possess, the Singularity will just not happen. Now, while it is true that various hardware measures seem to obey Moore's law, improvements in software are much less spectacular. (The word processor I use is the latest version of the program. It certainly has more features than the version I was using ten years ago, but I never *use* those features. Indeed, the program is probably slightly less useful to me than it was ten years ago; I persevere with it because everyone else uses it and I need to exchange documents with people. The program I am using to typeset this book, which is called TeX, is a wonderful piece of software whose creator froze development on the program several years ago.[173] While there is some progress in the worldwide TeX community toward an even better typesetting program, progress is much slower than would be the case if Moore's law were in operation. Of course, the kind of software required to create the "intelligence explosion" has nothing to do with word processors or typesetting programs. But the point is the same: advances in software and in software methodologies come at a much slower rate. We simply may not be smart enough to generate the software that will lead to a Singularity.) Perhaps we will see a future in which incredibly powerful machines do amazing things — but without self-awareness; surely this is at least as plausible as a future that contains a Singularity.

Even if a Singularity is inevitable, I fail to see how it explains the Fermi paradox. We can ask, as Fermi might: where are the super-intelligences? The motives and goals of a super-intelligent post-Singularity creature may be unknowable to us — but so, presumably, would the motives and goals of any "traditional" K3 civilizations that may exist. Yet we are happy to think about how to detect such K3 civilizations. (In fact, we may have more chance of understanding the post-Singularity beings on Earth than we would of understanding extraterrestrials, because in some sense those entities would *be* us. We would, in some sense, have created them and possibly imprinted upon them certain values.) Even if we are unable to understand or communicate with super-intelligent entities, it does not follow that those entities must disengage with the rest of the physical Universe. A super-intelligence must, like us, obey the laws of physics; and presumably it would make rational economic decisions. So the same logic that suggests that an advanced technological civilization would quickly colonize the Galaxy leads us to conclude that a super-intelligence would also colonize the Galaxy — except it would do so more quickly and efficiently than "normal" biological life-forms.

Even if they choose not to colonize, even if post-Singularity entities transcend our understanding of reality — they go off into other dimensions (page 122) or spend their time creating the child universes that Harrison proposed (page 59), or engage in any activity bar exploration of our Universe — there would be non-augmented, normal-intelligence beings left behind. In our case, I feel most of mankind would choose not take part in the Singularity. But it does not follow that we would become extinct. Unless the super-intelligences felt they had to destroy us (why would they bother?), we could go on living as we have always done. We might bear the same relation to the super-intelligent beings as bacteria do to us — but so what? Two billion years ago bacteria were the dominant form of life on Earth, and by many measures (species longevity, total biomass, ability to withstand global catastrophe, and so on) they still are. The existence of humans simply does not affect bacteria. In the same way, the existence of super-intelligent beings need not necessarily affect humanity; they could do their weird stuff, and we could continue doing ours. And the existence of super-intelligent beings does not affect our ability to communicate with like-minded ETCs.

To my mind, the existence of a Singularity does not explain the Fermi paradox. It exacerbates it!

SOLUTION 29: CLOUDY SKIES ARE COMMON

The long night had come again.
Isaac Asimov, Nightfall

Whenever polls are taken of such things, Asimov's *Nightfall* is routinely voted as the greatest piece of SF below novel length. It tells the story of scientists on Lagash, a planet in a system of six stars. In reality, the chaotic orbit of Lagash would not allow the development of advanced life-forms. For the sake of the story, however, Asimov postulates that intelligent, technically advanced creatures have developed on the planet. Physicists there have recently discovered the law of universal gravitation, so they can predict the position of any of the six suns of Lagash. Their new knowledge also enables them to deduce the existence of a moon orbiting Lagash. The presence of the moon has to be deduced because it is not visible: having six suns means darkness never falls on Lagash. The planet never has night. *Nightfall* describes what happens on Lagash when a rare alignment of the moon and the six stars produces an eclipse, and the beings of Lagash for the first time see the night sky. It is a wonderful story.[174]

The astronomers on Lagash would find it difficult to develop what we call astronomy. Since light from their six suns drowns out light from any other astronomical body, they could not know of the existence of planets or stars. Historically, on Earth, the development of physical science depended critically upon having planets whose orbits scientists tried to explain. Without a clear view of the skies, how could Lagash astronomers possibly develop an understanding of the physical Universe or of their place in it? They could be our superiors in terms of intelligence, they might develop a technology beyond our own, but they would not attempt to contact us because they would not know of or even suspect our existence.

Although the situation in *Nightfall* is unlikely, one can think of many cases where the physical environment of an ETC would prevent them ever developing the notion that beings exist on other worlds. What if, as one philosopher asked, cloudy skies are common? No matter how intelligent the species, no matter how good its technology, those beings might never develop an understanding of the Universe beyond their planet. Interstellar communication would not take place. Perhaps there are thousands of ETCs out there — but they are behind cloud cover, or stuck near the Galactic center where the sky is eternally bright, or in any of a hundred environments that would render astronomy difficult. Does this explain the paradox?

This idea has made for some of the greatest SF stories, but it is difficult to accept it as an explanation of the Fermi paradox. As we shall see later, we do not know how many habitable planets exist — but we do know that the number is probably large. It is inconceivable that Earth is the *only* planetary environment with a clear view of the skies!

SOLUTION 30: INFINITELY MANY ETCS EXIST BUT ONLY ONE WITHIN OUR PARTICLE HORIZON: US

We all live under the same sky, but we don't all have the same horizon.
Konrad Adenauer

Michael Hart has an interesting way of considering the paradox that he has done so much to promote. To fully appreciate his argument, we have to understand the notion of a particle horizon.[175]

A particle horizon is easiest to explain in a static Universe. (The Universe is, of course, not static. It began in the Big Bang, has been expanding ever since, and recent findings suggest that it will expand for eternity. Taking into account the expansion of the Universe makes a discussion of particle horizons rather subtle. Fortunately, nothing is lost if we discuss the idea in terms of a static Universe.) Imagine, then, a Universe that is infinite in extent and throughout which galaxies are uniformly distributed. Furthermore, this model Universe came into existence about 15 billion years in the past; perhaps the galaxies already existed, and some supreme intelligence "threw the switch" and turned on all the stars at precisely the same instant. What would such a Universe look like to an observer on an Earth-like planet, some 15 billion years after this creation event? Would the night sky be blindingly bright, the result of light reaching the planet from the infinite number of galaxies? Surprisingly (at least to those unfamiliar with Olbers' paradox), this infinite static Universe would look similar to our own Universe. The point to remember is that nothing can travel faster than light. So no influence — no light, no gravitational waves, *nothing* — could have reached the observer from regions more distant than 15 billion light years. This distance — the distance to the particle horizon — is the effective size of the observable Universe. Nothing from beyond the horizon has had time to reach the observer.

Hart makes the following argument. First, suppose our Universe is infinite. Since the Universe began some 15 billion years ago, however, the size of the *observable* Universe is given by the distance to the particle horizon. Second, suppose *biogenesis* — the development of life from non-living material — is an exceedingly rare occurrence. (We shall discuss the problem of biogenesis in more detail later, but at this point it is sufficient to say that Hart believes the probability of generating the characteristic molecules of life through the random shuffling of simpler molecules is exceptionally small. Most biologists think biogenesis must be common, because life arose so quickly on Earth; nevertheless, our knowledge of these things is so sketchy that Hart's contention cannot be ruled out.) It follows that in an infinite Universe there will necessarily be an infinite number of planets with life, but within any given particle horizon there might only be one planet with life. According to this argument, there is a sense in which there is nothing particularly special about Earth: in an infinite Universe there will be an infinite number of other Earths teeming with life. But within *our* particle horizon — within *our* observable Universe — only Earth spontaneously gave rise to life.

As Hart points out, his idea can be falsified quite easily. For example, extraterrestrials could visit Earth; or SETI might succeed and detect signals; or astrobiologists might prove that life arose spontaneously on Mars and independently of Earth. Any of these developments would disprove the notion of biogenesis being a rare, once-in-a-Universe event. In the absence of these developments, though, Hart argues that the Fermi paradox leads to a chilling conclusion: we are the only civilization within our particle horizon. Although the Universe contains an infinite number of advanced civilizations, for all practical purposes we are alone.

*   *   *

The conclusion that we are alone in the Universe — the third class of solution to the Fermi paradox — is the subject of the next chapter.

5

They Do Not Exist

The final class of solutions to the Fermi paradox is that "they" — extra-terrestrial civilizations advanced enough for us to communicate with them — do not exist.

Within this class of solutions, one can discern different approaches to Fermi's question. Ultimately, though, these solutions depend upon making one or more of the terms in the Drake equation tiny. If a single term is close to zero, or else if several of the terms are small, the effect is the same: when all the terms are multiplied together, the result is $N = 1$. The only technologically advanced civilization in the Galaxy, and perhaps the whole Universe, is our own.

Recently, Peter Ward and Don Brownlee, scientists at the University of Washington, wrote a stimulating and thought-provoking book called *Rare Earth*.[176] They presented a coherent argument about why complex life may be an unusual phenomenon. (Strangely, they make no mention of the Fermi paradox.) In this chapter I will discuss several of the ideas made in *Rare Earth*. Since each of these ideas has been proposed individually as a res-olution to the Fermi paradox, I discuss them individually. However, they could equally have been grouped as a single "Rare Earth" solution to the paradox.

Advanced ETCs may not exist because of a lack of suitable environ-ments: Earth-like planets may be rare. But perhaps they do not exist be-cause life itself is a rare phenomenon; perhaps the emergence of life from non-living material is an almost miraculous fluke, or perhaps the evolution of *complex* life-forms is unlikely to occur. I will discuss several solutions established on these ideas, but it is worth bearing in mind that the discussions will contain a major limitation: I will assume throughout that naturally occurring life is carbon-based and requires water as a solvent. Some scientists have argued that other chemicals, notably silicon, could be used instead of carbon; some have even argued that other solvents, perhaps methane, could be used in place of water. Personally — and this may be a failure of imagination on my part — I find it difficult to conceive of a biochemistry that does not feature water or carbon. Water in particular I am sure is necessary for life. Find water, and you have a chance of finding life. If you believe life can take quite different forms — perhaps as persistent patterns in plasma clouds, or as information-carrying whorls in viscous fluids, or whatever — then the solutions I present here will seem narrow-minded.[177]

We may later discover that many of the proposed solutions in this chapter stemmed from a lack of scientific imagination. But we are in the difficult position of trying to generalize from a single instance — as far as we know, Earth is the only planet with life. It is dangerous to draw conclusions from a sample size of one, but in this case what else can we do? Inevitably we will be influenced — perhaps biased is a better word — by those factors that seem necessary for our continued existence. We are bound by the Weak Anthropic Principle (WAP), which states that what we can observe must be restricted by the conditions necessary for our presence as observers. Since it is impossible to avoid the WAP in a discussion of the Fermi paradox, it makes sense to begin this part of the book with a solution based upon anthropic reasoning. Anthropic solutions are rather abstract; later solutions will be based on more concrete proposals.

SOLUTION 31:   THE UNIVERSE IS HERE FOR US

Man is the measure of all things.
Protagoras

A remarkable argument, which predates Hart's seminal analysis of the Fermi paradox, suggests mankind is probably alone. The argument relies on there being a number of "difficult steps" on the road to the development of a technologically advanced civilization. Examples of potentially "difficult steps" that we will discuss later include the genesis of life, the evolution of multicellular animals and the development of symbolic language. The precise details of the steps are, however, unimportant here. The argument simply requires there to be a number of critical yet unlikely steps on the road to intelligence. (The eminent evolutionary biologist Ernst Mayr once listed over a dozen of these "difficult" steps.[178] Other scientists have suggested the number might be even greater, particularly if certain physical and astronomical coincidences are added to the list.) Some of the evolutionary steps we call "difficult" may not be hurdles at all. We think of a particular evolutionary step as difficult if it occurred only once in Earth's history; but some steps probably *could* be taken only once — the competition they stimulated would have made a second occurrence redundant. On the other hand, some steps may have been genuinely improbable. For example, if a particular critical step required several otherwise worthless mutations to take place at the same time, then it makes sense to regard the step as a fluke.

Now consider a remarkable coincidence, which lies at the heart of the argument below.

On the one hand, the lifetime of our Sun is about 10 billion years. The period over which it can sustain life-bearing planets may be less than this — some astronomers believe the future evolution of the Sun will cause Earth to become uninhabitable in another 1 or 2 billion years, so the "useful" lifetime of the Sun could be as little as 6 or 7 billion years. On the other hand, *Homo sapiens* arrived on the scene when the Sun was about 4.5 billion years old. These two timescales — the lifetime of the Sun and the time for the emergence of intelligent life to appear around the Sun — are certainly within a factor of 2 of each other, and could even be within a factor of 1.3 of each other. The near equality of these timescales is really quite incredible. The two timescales are determined by factors that, either individually or in combination, would seem to have nothing to do with one another. The Sun's lifetime is determined by a combination of gravitational and nuclear factors, while a combination of chemical, biological and evolutionary factors determines the time of emergence of intelligent life. We live in a Universe in which timescales span a vast range: many subatomic processes occur on timescales as short as 10^{-10} seconds, while many astronomical processes occur on timescales as long as 10^{15} seconds. The typical times of certain other processes are even more extreme. The likelihood that two completely independent timescales have almost the same value is remote. How can we explain this observation without resorting to coincidence?

One solution would be if the evolutionary timescale is much *smaller* than 4.5 billion years. Suppose the typical time for the evolution of intelligent life on an Earth-like planet is just 1 million years. The coincidence of timescales would lessen — but at the expense of making the probability of mankind's recent emergence vanishingly small. After all, if we *could* have emerged just 1 million years after the Earth cooled, then why do we not observe the Earth to be 1 million years old? At the very least, why do we not observe it to be 2 million years old, or 3, or 4? Why did it take 4.5 billion years for us to appear? This is not a good solution.

The other solution requires the evolutionary timescale to be much *longer* than 4.5 billion years. This accords with Mayr's suggestion of a number of difficult steps in the development of intelligence — "difficult" in this sense meaning that, on a given viable planet, the typical time for a step to occur is long (perhaps longer than the present age of the Universe). If several difficult steps must be taken, then we would not expect to be here at all!

Most people, upon hearing this second solution, dismiss it on the same grounds as the first solution: the probability of mankind emerging recently is small. But the two situations are not equivalent.

Consider the ensemble of all possible universes. (Whether you consider these universes as somehow "real" or as some sort of mathematical idealization is up to you.) In some universes, unlikely things will occur; a chain of improbable events will happen. In some universes, due to the blind workings of chance, the set of difficult steps leading to intelligence will happen. *And it is precisely such a universe an intelligent species will observe — with themselves in it.* In other words, we can ignore the possible universes in which we do not exist — since by definition they do not exist for us. We *must* observe those universes in which the difficult steps have occurred and led to us. Now we can ask: Of all the universes that exist for us, when are we most likely to emerge, given that we can only emerge some time in the 10-billion-year total lifetime of the Sun? (Or, if it happens to be the case, the 6- to 7-billion-year *useful* lifetime of the Sun?) A simple calculation shows that if there are 12 difficult steps, then the most likely time of emergence is after 94% of the star's available lifetime has passed.

Our observations seem to be consistent with the results of this simple calculation. If the Sun were able to sustain life on Earth for 10 billion years, then mankind emerged after roughly 50% of the time available had elapsed. However, if, as some astronomers believe, the Sun can sustain life for only another billion years or so, then mankind emerged after roughly 83% of the time available. This is impressively close to the expected time of arrival.

The Most Probable Time of Emergence of a Communicating Civilization

Suppose there are n difficult steps on the road to the development of a civilization capable of interstellar communication. And suppose these steps must take place over the lifetime L (in years) of a star. A straightforward calculation shows that the most probable time of emergence of a communicating civilization is given by the expression $L/(2^{1/n})$. If there are a dozen difficult steps, so $n = 12$, then the most probable time of emergence is $0.94L$. The calculation does not determine exactly when an intelligent species will emerge; just that the median time of emergence, if there are 12 difficult steps to negotiate, is 94% of the star's lifetime.

Finally, we come to the key point. Merely because we have selected universes in which *we* exist (and how could we select any other type of universe?), we cannot infer that *other* intelligent species exist. We *have* to be here because we observe ourselves to be here; but the existence of aliens must contend with probabilities, and the odds are not good. Another calculation makes this clear. If there are a dozen difficult steps to negotiate on the road to high intelligence, then even under generous assumptions there is only one chance in a million billion of there being another intelligent species in our whole Universe. No wonder we do not observe them!

The Number of Intelligent Species in Our Universe

Suppose there are n difficult steps on the road to intelligence and each step typically requires d years to occur. Furthermore, suppose there are p viable planets, each of which could have supported life for t years. The number of intelligent species out there is given by the expression $p \times [t/(n \times d)]^n$. Let us be generous and suppose every star in every galaxy possesses a viable planet; so $p \approx 10^{22}$. Let us be even more generous and suppose every planet has been viable for about the age of the Universe, so $t \approx 10^{10}$ years. However, d must be long: that, after all, is what makes the step difficult. So let us suppose $d \approx 10^{12}$ years — 100 times the age of the Universe. Finally, let us suppose as before there are a dozen difficult steps, so $n = 12$. If we plug these numbers into the expression above, we find the number of intelligent species out there is 10^{-15}.

145

Chapter 5

This type of argument for the non-existence of ETCs was first presented by Brandon Carter.[179] It is known as an *anthropic* argument. (We have met anthropic ideas before in this book: the doomsday argument of Gott and Hart's suggestion regarding the improbability of life's genesis have anthropic overtones. We will meet other examples.) Carter's use of the term "anthropic" was perhaps unfortunate, since it implies mankind is somehow necessary. All that is needed for the argument to work is that intelligent observers — *any* intelligent observers — self-select their Universe. However, in this Universe it is *we* who make the observations.

The status of anthropic reasoning in science is contentious. Some view it as an abdication of the scientists' responsibility to provide explanations. For example, Smolin's idea of natural selection acting on whole universes (see page 57) is an attempt to move away from anthropic reasoning. Nevertheless, many respectable scientists have employed anthropic ideas in an attempt to explain several features of the Universe that seem to be "just right" for the evolution of life; if certain physical constants possessed only slightly different values, then we would not be here. Stars would not shine, or the Universe would have collapsed in on itself in a fraction of a second, or heavy elements could not form, and so on. The fact of our existence can perhaps, in some way, make sense of these observations. (But I think one can equally argue that these "explanations" are essentially trivial.)

There are several types of anthropic reasoning, corresponding to several anthropic principles each with different shades of meaning. According to Carter, the weak anthropic principle (WAP) is that "what we can expect to observe must be restricted by the conditions necessary for our presence as observers." The WAP seems almost tautologous. The strong anthropic principle (SAP), on the other hand, is more contentious: "the Universe (and hence the fundamental parameters on which it depends) must be such as to admit the creation of observers within it at some stage." Barrow and Tipler, in a classic book, also discuss the final anthropic principle (FAP), which they define as "intelligent information-processing must come into existence in the Universe and, once it comes into existence, it will never die out."[180] The mathematician Martin Gardner, in his inimitable way, calls this latter version the completely ridiculous anthropic principle (CRAP).

It is interesting that Tipler expanded upon the notion of the FAP in a book entitled *The Physics of Immortality*.[181] He considered the far future of the Universe, and was lead to a concept not unlike Teilhard de Chardin's Omega Point. His work showed that, if the Universe collapses in a Big Crunch, then a future intelligence would find it possible to perform an infinite number of computations. Every being who ever lived could be "resurrected" as a computational simulation. According to his interpretation of the FAP, the Universe must be such that it allows this infinite amount of

146

information processing. Now, although Tipler's ideas were attacked as being altogether too speculative (and too overtly religious), his hypothesis at least had the virtue of being falsifiable. He made a definite, testable prediction: the Universe is closed and will collapse on itself. Recent observations, however, seem to indicate that the Universe is not only open, it is expanding *more rapidly* as it ages. Tipler, it appears, was wrong; his interpretation of the FAP seems disproven. Perhaps one day soon we will discover signals from extraterrestrials, or even receive a visit from them. Such a discovery would cast in doubt the WAP and SAP. I leave the reader to decide whether such a discovery is probable.

SOLUTION 32: LIFE CAN HAVE EMERGED ONLY RECENTLY

To everything there is a season,
and a time to every purpose under heaven.
Ecclesiastes 3:1

The astronomer Mario Livio takes issue with the notion that the timescale for the evolution of intelligent life is completely independent of the main sequence lifetime of a star. If the two timescales were related in a particular way — if the evolutionary timescale increases as a star's lifetime increases — then we would *expect* to observe the two timescales as roughly equal. Carter's gloomy conclusion about the non-existence of ETCs would then not follow. But how can the lifetime of a star influence the timescale of biological evolution?[182]

Livio considers a simple model of how a planetary atmosphere like Earth's develops to the stage where it can support life. It is *not* a serious model of atmospheric development; rather, it is meant to demonstrate a possible link between stellar lifetimes and the timescale for biological evolution.

In his model, Livio identifies two key phases in the development of a life-supporting atmosphere. The first involves the release of oxygen from the photodissociation of water vapor. On Earth, this phase lasted about 2.4 billion years and resulted in an atmosphere with oxygen levels at about 0.1% of present values. The duration of this phase depends upon the intensity of radiation emitted by the star in the wavelength region of 100 to 200 nm, because only this radiation leads to the dissociation of water vapor.

The second phase involves an increase in oxygen and ozone levels to about 10% of their present values. On Earth, this phase lasted about

1.6 billion years. Once oxygen and ozone levels were high enough, Earth's surface was shielded against ultraviolet (UV) radiation in the wavelength region of 200 to 300 nm. This shield was important because it protected two key ingredients of cellular life: nucleic acids and proteins. Nucleic acids absorb radiation strongly in the wavelength region of 260 to 270 nm, while proteins absorb radiation strongly in the wavelength region of 270 to 290 nm; radiation in the region of 200 to 300 nm is therefore lethal to cell activity. It is vital — at least for land life — that an atmosphere develops a protective layer for these wavelengths. And of the likely candidates of a planet's atmosphere, only ozone absorbs efficiently in the wavelength region of 200 to 300 nm: *a planet needs an ozone layer*. Livio argues that, as on Earth, the timescale for developing an ozone shield against UV radiation is roughly equivalent to the timescale for the development of life.

Different types of star emit different amounts of energy in the UV region. High-mass stars are hotter than low-mass stars and emit more UV radiation, but they have shorter lifespans. So for a given planetary size and orbit, the timescale for developing an ozone layer depends upon the type of radiation emitted by the star, and thus on the star's lifetime. Following a detailed calculation, Livio argues that the time needed for intelligent life to emerge increases almost as the square of the stellar lifetime. If such a relation holds, then we are *likely* to observe intelligent species to emerge on a timescale comparable to the main-sequence lifetime of a star.

The purpose of Livio's model, to repeat, is simply to show whether a relationship *possibly* exists between the timescale for biological evolution and stellar lifetimes. Even with this proviso, one can still disagree with parts of Livio's argument. For example, his model involves a *necessary* condition for land life to evolve (namely, the development of an ozone layer); but this is not a *sufficient* condition. There are many other steps on the road to the evolution of intelligent life, so even if there is a link between stellar lifetimes and the timescale for biological evolution, it may be a minor factor. Nevertheless, encouraged by the discovery of a link between these timescales and the possibility therefore that the existence of ETCs is not ruled out, Livio asks the following question: in the history of the Universe, when is a likely time for ETCs to emerge?

If life on Earth is typical of life elsewhere, then most life-forms will be carbon-based. Livio therefore suggests that the emergence of ETCs will coincide with the peak in the cosmic production of carbon. And this is something we can calculate.

The main producers of cosmic carbon are planetary nebulae, which occur at the end of the red-giant phase of average-mass stars. Planetary nebulae shed their outer layers into the interstellar medium, and the material is recycled to form later generations of stars and planets. Since astronomers

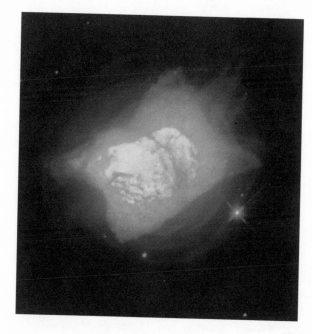

FIGURE 44   *The planetary nebula NGC 7027. Objects like this produce much of the carbon we observe in the Universe.*

believe they know the historical rate of star formation (it was higher in the past than it is now, with a peak about 7 billion years ago) and they know the relevant details of stellar evolution, they can calculate the rate at which planetary nebulae formed in the past — and thus the rate of cosmic carbon production. According to Livio's calculations, the rate of planetary nebula formation peaked a little less than 7 billion years ago. From this, he argues we might expect carbon-based life to have started when the Universe was about 6 billion years old. Since the time required for advanced ETCs to evolve is a significant fraction of a stellar lifetime, we would expect ETCs to develop only when the Universe was about 10 billion years old. If this is the case, then ETCs cannot be more than about 3 billion years older than us.

Livio's conclusion has been proposed by others as a resolution of the Fermi paradox. They suggest life could have emerged only relatively recently on a cosmic scale. There are presently no ETCs capable of interstellar travel or communication because, like us, they have had insufficient time to develop. Perhaps in the future the Galaxy will be aswarm with interstellar commerce and travel and gossip. For now, though, all is silence.

But even if Livio's conclusion is correct, and there are no ETCs more than 3 billion years in advance of us, I fail to see how it solves the Fermi paradox. An ETC that is 3 billion years older than us has had *plenty* of time to colonize the Galaxy; it has had *plenty* of time to announce its presence to the Universe. (In the Universal Year, ETCs could have reached our present level of technology at about October 1; they thus have 3 months to colonize the Galaxy — a process we can measure in hours on this scale. They have had time enough to get here.) Unless it can be shown that intelligence is only coming into existence *now*, and thus life on Earth is among the most advanced in the Galaxy, the arguments do not really address the main thrust of the paradox.

SOLUTION 33:   PLANETARY SYSTEMS ARE RARE

A time will come when men will stretch out their eyes.
They should see planets like our Earth.
Christopher Wren, *Inaugural Lecture as Professor of Astronomy, Gresham College*

Anthropic arguments are rather abstract. There have been many more tangible suggestions as to why ETCs might not exist. For example, perhaps there is no place for them to develop.

A common assumption is that complex life requires a planet — preferably Earth-like — on which to originate and evolve. A technologically advanced species may one day decide to move away from planet dwelling, of course, but the evolutionary ancestors of those species must have began as planet dwellers. (Some SF writers have explored the possibility of life evolving in more exotic locales, including the surface of a neutron star and a gas ring around a neutron star.[183] Although these fictional accounts are often surprisingly plausible, it remains far easier to imagine such possibilities than it is to demonstrate convincingly and in detail how complex life could originate and evolve anywhere other than on a planet.) When Sagan arrived at his figure of 1 million ETCs in the Galaxy, he assumed there might be as many as 10 planets per star. But perhaps planetary systems are rare, and the f_p term in the Drake equation is small. If f_p were small enough, this alone could explain the Fermi paradox.

*   *   *

Not so long ago, astronomers were still not certain how planets formed. There were two competing scenarios. In the first, a planetary system like ours was pictured as forming in a catastrophic event. In the second, planetary systems were thought to condense out of nebulae.[184]

The nebular hypothesis feels like the most "natural" explanation, but it seems to possess a fatal flaw. If the Sun, for example, formed from the collapse of a rotating cloud of dust and gas, then calculations show that it should now rotate *extremely* quickly. The Sun should contain most of the angular momentum of the Solar System. And yet it does not. In fact, the Sun rotates rather sedately — its equatorial regions rotate once in about

24 days, while its polar regions rotate once in about 30 days. This observation led many astronomers to prefer models of planetary formation based on catastrophic events. The most popular model had a star almost colliding with the Sun; tidal effects pulled a gaseous filament from the Sun, and the filament later broke up and condensed to form the planets.[185]

If planets really did form in stellar collisions, then the outlook for finding ETCs would be bleak. The density of stars in space is quite low, so collisions would be infrequent; one early estimate put the number of planetary systems formed in this way at just ten per galaxy! In a lecture in 1923, James Jeans said: "Astronomy does not know whether or not life is important in the scheme of things, but she begins to whisper that life must necessarily be somewhat rare." Jeans clearly thought he knew the resolution of the paradox, and the paradox had not yet been formulated.

However, the nebular hypothesis never went away. Theories of planetary formation based upon collisions also possessed problems. The collision theory could not explain many observed properties of our Solar System. Furthermore, the major difficulty with the nebular hypothesis — namely, explaining how the bulk of the angular momentum of the Solar System resides in the planets — was eventually resolved. It happens that the young Sun *did* rotate at high speed, but the rotation generated a strong magnetic field. Magnetic lines of force stuck out into the solar nebula, like spokes from a hub, and dragged the gas around with it. This "magnetic braking" effect slowed the Sun, and transferred angular momentum to the gaseous disk. Astronomers observe direct evidence for this: young stars rotate up to 100 times as quickly as our Sun, whereas old stars rotate more sedately. Few astronomers now doubt that the planets in our Solar System formed when small planetesimals condensed out of a disk-shaped cloud of dust and gas; in gentle collisions, these planetesimals stuck together and gradually formed the planets we see today. If this theory is correct, then the same process should occur around other stars. Planets should be common, as Sagan believed.

Astronomers have even photographed protoplanetary disks, which has lent credence to their theory of planetary formation. But it is one thing to photograph a disk of gas that *may* one day become a planetary system; it is quite another to photograph a planet.

It is not feasible, at least at present, to *see* planets around distant stars. Planets shine only by reflected light, so taking a photograph of an extrasolar planet is rather like trying to observe the light of a firefly next to a thermonuclear explosion. Recent advances in observational astronomy, however, have made it possible to infer the existence of planets around other stars by the gravitational pull they exert on their parent stars. The gravitational attraction of a large planet on a star causes the path of the star

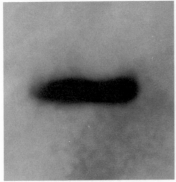

FIGURE 45 A protoplanetary disk.

to "wobble." By measuring the wobble, astronomers can not only determine the mass of the planet but also its distance from the star. The first planet detected by this technique was only found in the mid-1990s; but the technique is so successful that there are already more than 60 known extrasolar planets (the precise number depends on how you choose to define a planet), and more are being found each month.[186]

Clearly, then, it is simply wrong to attempt to explain the Fermi paradox by stating that the number of stars with planetary systems — and thus the total number of planets — is small. We now know of too many planetary systems to accept this argument.

And yet . . . so far astronomers have found only giant planets — planets with a mass similar to that of Jupiter. This is not surprising: using the technique described above, astronomers *can* only find giant planets. But of the stars tested to date, less than 10% of them have detectable planets. This could be because detectable Jupiter-sized planets are relatively rare — but it *could* mean planets in general are quite rare; certainly, not every star has a planetary system. Furthermore, as we shall discuss in later sections, the Jupiter-sized planets found to date tend to be either extremely close to their sun or, if they orbit at larger distances, they have extremely elliptical orbits. In either case there is little chance of a habitable Earth-like planet existing in these systems. A "Jupiter" close to its star will destroy rocky Earth-like planets, while a "Jupiter" in an elliptical orbit will disrupt the orbits of smaller planets, either casting them out into space or throwing them into the central star.

Personally, I believe the f_p term in the Drake equation will turn out to be smaller than the early optimists believed — but by itself it will still be *far* too high to permit a resolution of the Fermi paradox. Fortunately, this will soon cease to be a matter of belief; rapid advances in observational astronomy mean that within a few years we should have a clear understanding of the number and type of extrasolar planetary systems.

SOLUTION 34: WE ARE THE FIRST

'tis not the king's stamp can make the metal better or heavier.
William Wycherly, *The Plaindealer*

The biochemistry of terrestrial organisms — and the biochemistry of any extraterrestrial organisms we can plausibly imagine — depends crucially on six elements: sulfur (S), phosphorus (P), oxygen (O), nitrogen (N), carbon (C) and hydrogen (H). To an astronomer, any elements heavier than hydrogen and helium are known as metals. (The *metallicity* of a star, then, refers to the amount of heavier elements it possesses.) So in astronomical language, life depends upon the five metals SPONC. Soon after the Big Bang, the Universe contained essentially only hydrogen and helium (in the ratio 75% to 25%). The Big Bang would have produced small amounts of lithium, and even smaller traces of beryllium and boron. But that was it: none of the metals required by life were there at the beginning. One of the key findings of modern astronomy is that the heavier elements like SPONC were cooked in nuclear reactions inside stars, and became part of the interstellar medium only when stars reached the end of their energy-producing life. As time goes by, the concentration of metals in the Universe slowly increases.

One resolution of the paradox — often proposed and similar in spirit to Livio's suggestion — is that the heavier elements only recently became sufficiently concentrated in the interstellar medium to allow life to form. Planets around older stars, it is suggested, lack the metals SPONC. Only around quite young stars — stars like the Sun — can life arise. So mankind would inevitably be among the first civilizations, perhaps *the* first, to arise.

*   *   *

Like many of the proposed solutions we have discussed, the suggestion that the chemical enrichment of the Galaxy resolves the Fermi paradox *by itself* is too strong. It may be a *factor* in the final explanation, but for two reasons it is unlikely to stand alone as a resolution.

First, we do not know what metallicity is required of a star for it to possess viable planets. Would an abundance of heavy elements one third that of the Sun suffice? One quarter? One tenth? We simply do not know. So far, no planets have been found around any star that has a metallicity less than 40% that of the Sun, but these observations are in their infancy. If life can develop on planets possessing a much smaller abundance of heavy elements, then very old stars could be home to life.

Second, the metallicity of stars differs between the four *stellar populations*. Some types of star might be ancient and yet be metal-rich. The four stellar populations consist of the thin disk stars, the thick disk stars, the halo stars and the bulge stars. The halo stars, which form a spherical system around the center of the Galaxy, are old stars. They typically have a metallicity about 1% that of the Sun. Such stars are unlikely to possess planets. The bulge at the center of the Galaxy is old, and yet some of the stars are very rich in metals. However, bulge stars orbit within a few thousand light years of the Galactic center, which is a violently energetic environment. Whether complex life-forms can exist in such an environment is debatable, and too high a metallicity can itself be a problem, so it is safest to ignore bulge stars in these discussions. The thick disk consists of stars that stay reasonably close to the plane of the Galaxy. (But not *too* close; stars can move a few thousand light years above or below the plane — hence the term "thick" disk.) Such stars are old, and their metallicity is generally 25% that of the Sun. Finally, the thin disk stars, which stay within 1000 light years of the plane of the Galaxy, are the interesting ones from our point of view. Not only is the Sun a member of the thin disk population — so are 96% of its neighbors. These stars have a variety of ages, ranging from objects that are 10 billion years old to stars that have formed only recently. Similarly, the metallicities of the thin disk stars vary: some have less than 1% of the metallicity of the Sun and are poor candidates for life, but some have *three times* the Sun's metallicity. So the situation is more complicated than at first appears. It seems, though, that within all this variability there are many stars much older than the Sun but with the same abundance of heavy elements.

Consider, for example, 47 Ursae Majoris — a thin disk star only slightly more massive and only slightly hotter than our Sun. By coincidence, on the day I am writing this section astronomers have announced the discovery of a second Jupiter-sized planet orbiting the star.[187] The discovery of 47 UMaj c (as the planet is tentatively called, and presumably will continue to be called until astronomers can decide upon a better nomenclature for extrasolar planets) is interesting for two reasons. First, 47 UMaj c is orbiting in a nearly circular orbit around the star, as is its companion 47 UMaj b. This planetary system is the first that might turn out to be like our own, in that the planetary orbits have low eccentricities and the Jupiter-sized planets are at a respectable distance from the star. (So arguing that the scarcity of "good Jupiters" resolves the Fermi paradox, as we do on page 160, may turn out to be wrong.) Second, 47 Ursae Majoris is 2.5 billion years older than the Sun and yet it has essentially the same chemical composition. Thus any Earth-like planet orbiting this star could in principle have given birth to life some 2.5 billion years ago; an ETC on that planet could be in advance of us by 2.5 billion years. This corresponds to almost $2\frac{1}{2}$ months in the Universal Year — much longer than the colonization time of the Galaxy.

(It must be stressed that astronomers do not know whether small, rocky planets exist in the inner planetary system of 47 Ursae Majoris. Our present techniques simply cannot detect such objects. Nevertheless, this planetary system is undoubtedly the most similar to our own. The ratio of the masses of 47 UMaj b to 47 UMaj c is 3.3 to 1, which is the same as the ratio of masses of Jupiter and Saturn. The ratio of their average distances from 47 Ursae Majoris is the same as the ratio of the average distances of Jupiter and Saturn from the Sun. Finally, since the observations suggest that there can be no further giant planets orbiting closer to the star than 47 UMaj b, there would seem to be "room" for Earth-like planets to exist. Unfortunately, numerical simulations suggest there are probably no Earths there: 47 UMaj b and 47 UMaj c orbit closer to their parent star than Jupiter and Saturn orbit the Sun, so their gravitational influence would disrupt the formation of terrestrial planets at the correct distance from the star. But one can dream.)

Regardless of whether 47 Ursae Majoris turns out to possess terrestrial planets, the fact remains it is a Sun-like star, it possesses the same chemical abundances as the Sun, and it has planets. The star is a neighbor — less than 50 light years away from us. Yet it is 2.5 billion years older than the Sun. If such stars are in our backyard, there must be many of them in the Galaxy. Perhaps the number of stars that are suitable for hosting life-bearing planets is much smaller than previously thought, but the suggestion that the Sun is among the earliest generation of stars that can give rise to life seems to be untenable.

There is, however, a further point to make. Although *our* Galaxy may possess millions of old stars with sufficient metals to sustain life, the same is not necessarily true of all galaxies. Elliptical galaxies, for example, generally contain metal-poor stars, and they are not the best place to look for life. Small irregular galaxies, too, are unlikely to be home to life as we know it. Furthermore, globular clusters (collections of millions of stars that orbit larger galaxies) are also metal-poor regions. Although Earth's first dedicated interstellar radio transmission was to the globular cluster M13 (see page 111), the signal is thus unlikely to cross an Earth-like planet there. The chemical enrichment of galaxies may help explain why we do not see K3 civilizations: there might be far fewer galaxies that are suitable for life than we at first expect.

SOLUTION 35: ROCKY PLANETS ARE RARE

Here's metal more attractive.
William Shakespeare, *Hamlet*, Act 3, Scene 2

As far as we know, the only surviving witnesses to the birth of the Solar System are a group of metal-rich meteorites called *chondrites*. (Their name comes from the Greek word *chondros*, which means "grain" or "seed": it refers to the appearance of the many small spherical inclusions, known as *chondrules*, which occur within them. Chondrules are typically 1 to 2 mm in diameter, and are composed mainly of the silicate minerals olivine and pyroxene.) Using the known decay rates of various radioisotopes found in chondrites, we can calculate when these meteorites formed. The best estimates imply that chondrites formed about 4.56 billion years ago, which is the accepted age of the Solar System. Chondrites, it seemed, formed within the first few million years of the Solar System's history.[188]

FIGURE 46 *Chondrules are spherical inclusions of silicate in chondrites. Their origins remain a matter for debate. Chondrules are clearly visible in this cut surface of the AH 77278 chondrite. This specimen is 8 cm across.*

Chondrites occasionally fall to Earth, and when they do they are studied intensively. Indeed, chondrites have been studied for over two centuries, and much is known about their chemical and physical makeup. One thing remains mysterious, though: the precise nature of the chondrules.[189]

An embarrassing number of hypotheses have been presented to explain the enigma of chondrule formation. (A surfeit of hypotheses is a sure sign that we do not understand something. In the case of chondrule formation, this lack of understanding is not surprising. Not only did chondrules form a long, long time ago, but they appear in no other type of rock. Geologists have no other specimens with which to compare them.) The ideas range from the suggestion that chondrules are drops of lava ejected from extra-terrestrial volcanoes to the suggestion that they form when lightning discharges through dust balls. All we know for sure is that chondrules must have been flash-heated to temperatures above 1800 K, and then cooled quickly. One interpretation is that, about 4.5 billion years ago, a brief flash of heat spread through the Solar System.

In 1999, the Irish astronomers Brian McBreen and Lorraine Hanlon proposed a new theory of chondrule formation: they suggested a gamma-ray burster (GRB) might have been involved.[190] Suppose a GRB occurred within 300 light years of the nascent Solar System. It would have pumped enough energy into the protoplanetary ring of dust and gas to fuse up to 6×10^{26} kg of material (100 times the Earth's mass) into iron-rich droplets, which would quickly cool to form chondrules. The chondrules would then absorb gamma-rays and X-rays from the GRB.

If McBreen and Hanlon are correct, then the Solar System could be a rarity in possessing chondrules. They estimate that, on average, only 1 star in 1000 would be close enough to the burster for chondrule formation to occur. The significance is that the high-density chondrules may have settled quickly into the plane of the protoplanetary disk and aided the formation of the planets. In other words, planetary systems like our own — complete with rocky terrestrial planets — would be scarce. And, with only a small number of Earth-like planets on which to develop, ETCs might be vanishingly rare.

The idea that chondrule formation was initiated by a GRB is interesting. However, other suggestions seem to offer more plausible mechanisms for making chondrules. Furthermore, these other mechanisms do not imply that there is anything particularly special about our Solar System. So, as a solution to the Fermi paradox, this is not high among the list of contenders.

*   *   *

A discussion of metal-rich meteorites brings to mind a related solution to the Fermi paradox that surfaces occasionally: perhaps planets with workable lodes of metallic ores are rare. The reasoning is simple: if alien intelligences cannot find and work metal, then they will be unable to develop technology — and thus will be unable to construct the radio telescopes or starships that would enable them to contact us.

This solution has been well examined by several SF authors. One group of authors has dismissed the suggestion in thought-provoking stories. Even if Earth's surface composition *is* unusual among planets (see page 183 for one reason why this might be the case), they believe this does not necessarily mean that technology is impossible. The technology would inevitably be different from ours, but the results might be the same. (For example, perhaps extraterrestrials produce electricity using biological means rather than generators?) A different group of authors — either less imaginative or more realistic, depending upon one's point of view — argue that technology cannot develop without the materials we take for granted.

We shall return to the issue of technological progress in a later section. However, whether or not technology is possible in the absence of metals (and this is something we may never know), it seems perverse to attempt to resolve the Fermi paradox by supposing Earth is the *only* planet in the Galaxy with workable lodes of metallic ores. A scarcity of such planets may yet be another factor acting against the existence of ETCs, but surely this cannot by itself explain the silence of the Universe.

SOLUTION 36: CONTINUOUSLY HABITABLE ZONES ARE NARROW

Give me more love or more disdain; the torrid or the frozen zone.
Thomas Carew, *Mediocrity in Love Rejected*

Even if terrestrial planets form readily around stars, another condition must be met before life as we know it can survive for the billions of years needed for a technological civilization to develop. A terrestrial planet has to be in a system's *habitable zone* (HZ) before life can evolve.[191]

The key to life is water. In essence, the habitable zone around a star is the region in which a planet like Earth could support liquid water. The location of the inner edge of the HZ is set by the point at which a planet loses water due to the high temperatures close to a star. The outer edge of the HZ is set by the point at which water freezes.[192] Many scientists believe that the HZ for our Solar System extends from 0.95 AU to 1.37 AU. Venus, with a mean distance of 0.723 AU from the Sun, lies inside the inner edge of the habitable zone; Mars, with a mean distance of 1.524 AU from the Sun, lies outside the outer edge of the habitable zone. Only Earth lies in the right place. Nevertheless, the habitable zone is rather wide: if the HZ was the full story, one would expect most other systems to have planets in the zone. It is, of course, not the full story.

Michael Hart argued that the habitable zone around a star varies with time. Main-sequence stars become brighter and hotter as they grow older, so the HZ moves outward as a star ages. What is important, according to Hart, is the *continuously* habitable zone (CHZ).

Typically, the CHZ is defined as the region in which an Earth-like planet can support liquid water for 1 billion years — the timescale evolution presumably requires to develop complex life. In the case of the Solar System, the CHZ has existed for 4.5 billion years, and Earth has been fortunate enough to be precisely in the middle of the zone. Clearly, though, the CHZ must be narrower than the HZ. In 1979, Hart published the results of computer models that seemed to show that the CHZ is extremely narrow.[193] It is widest around G0 main-sequence stars (the Sun is a G2 star) and shrinks to zero at cool K1 stars and hot F7 stars. In all cases, though, the CHZ was narrower than 0.1 AU. For the Solar System, for example, he calculated an inner edge of the CHZ at 0.95 AU and an outer edge at 1.01 AU. With such a narrow CHZ, one would expect Earth-like planets — those that can support life over billions of years — to be much rarer than is commonly supposed.

While Hart's finding did not *prove* ETCs could not exist, it clearly had a bearing on the Fermi paradox. If the number of potential life-bearing planets is much smaller than most estimates suppose, then the number of potential ETCs out there must also be smaller. Depending upon the values of the other factors in the Drake equation, the total number of communicating civilizations might be reduced to one: us.

Recent calculations, however, employ more sophisticated models of the Earth's early atmosphere; they also take account of the recycling of CO_2 by plate tectonics, a phenomenon not known to Hart. The results are encouraging for those who would believe in the existence of ETCs (or at least in the existence of planetary homes for ETCs). Models developed by James Kasting and co-workers suggest that the 4.6-billion-year CHZ for our Solar System extends from 0.95 AU to 1.15 AU — larger than the range calculated by Hart.[194] Other scientists believe the Solar System's CHZ may be even wider. The CHZ around other stars, too, is wider than first thought.

So: how likely is it that a given planetary system will have a planet that lies within the CHZ? The answer depends both upon the type of star and the distribution of the planets in the system. If planets are distributed as they are in our Solar System — in other words, if the distances of planets from the central star follow the Titius–Bode law — then roughly the same number of planets will exist in the instantaneously habitable zones of all stellar types. However, planets around hot stars of type O, B and A will not long remain in a habitable zone, as the stars themselves evolve in luminosity too quickly. Planets around cold stars of type K and M are unlikely to be continuously habitable: the HZ in these systems lie close to the central star, and the planet will thus become tidally locked. (When a planet is tidally locked, one side of the planet always faces the heat of the star, while the other side always faces the cold of open space. This situation is presumably inimical to life.) Around stars not too different from the Sun, however, a planetary system, if it obeys the Titius–Bode law, has roughly a 50:50 chance of containing a planet in the CHZ.

If our current models of planetary formation, stellar evolution and long-term planetary atmospheric evolution are correct (and it must be admitted that there are places where scientists are surely uncertain on the details), then the conclusion seems to be that there are potentially millions of continuously habitable planets in the Galaxy. One caveat, though. We saw in an earlier section that only certain types of star have sufficient metallicity to possess terrestrial planets; and only certain parts of the Galaxy are sufficiently protected from the violence of the central regions. We may need to define a galactic habitable zone (GHZ) — which is an annulus containing perhaps only 20% of the stars in the Galaxy. For complex life to evolve, a CHZ must be within the GHZ — and this reduces the possibilities.[195]

SOLUTION 37: JUPITERS ARE RARE

What men are poets who can speak of Jupiter if he were like a man, but if he is an immense spinning sphere of methane and ammonia must be silent?
Richard Philips Feynman, *The Feynman Lectures on Physics*

Since the first discovery in 1995 of extrasolar planets, or *exoplanets*, astronomers have found 60+ more planets beyond our Solar System. Many of these are Jupiter-sized objects orbiting in nearly circular orbits close to the parent star. (Consider the planet orbiting Rho CrB, for example. Of all the exoplanets yet discovered it is closest in mass to Jupiter, being only 1% less massive than Jupiter. However, whereas Jupiter orbits the Sun at about 5.2 AU (an astronomical unit being the Earth–Sun distance, which is a convenient distance measure for planetary systems), the massive planet around Rho CrB has a nearly circular orbit at just 0.224 AU. This means it is far closer to its star than Mercury is to our Sun; Mercury orbits at 0.387 AU. It is not surprising that massive planets orbiting close to a star should have circular orbits: tidal forces from gravitational interaction with the star will cause the orbit to become circular even if the orbit began as an ellipse. Nor is it surprising that astronomers can detect large planets orbiting close to a star: our present techniques for detecting planets work best on precisely such objects. The surprise is that so many Jupiter-sized planets exist in orbits so close to a star. These planets should not exist at all!

Our theories of planetary formation imply that gas-giant planets like Jupiter cannot form within 3 AU of a star like our Sun. This limit is called the *snow line*. So what are these so-called "hot Jupiters" doing so far within the snow line? With some detective work we can rule out one possibility; namely, that these are not actually gas giants. The Doppler motions that enable astronomers to infer the existence of the planets also give us enough information to deduce their masses; and in a few cases, measurements of the parent star during transits enable us to estimate the diameters of the planets. These two pieces of information directly give us the planets' densities — and they are certainly gas giants. A second possibility — namely, that our models of planetary formation are wrong — cannot be dismissed. However, there is a lot of evidence to support the models, and there is nothing to replace them; so astronomers are loath to accept this possibility. Which leaves a third possibility: the planets formed outside the snow line and later migrated to their present positions close to their parent stars.

The orbital decay of Jupiter-type planets cannot happen once a planetary system is established, so we need not worry about a Jovian threat in our own Solar System. But the decay can happen early in the development of a planetary system. If a gas giant migrates from outside the snow line to an orbit close to the star, then the outlook for any inner terrestrial planets is bleak. Simulations show that smaller planets are forced into the star, or ejected from the planetary system altogether. Stars with "hot Jupiters" are unlikely to possess viable planets.

Not all exoplanets are hot Jupiters. Some of them are outside the snow line, where we would expect them to be. An example is the planet around Epsilon Eridani. (This is one of the closest Sun-like stars, and one that Frank Drake observed when he carried out the first search for extraterrestrial signals.) The planet, designated Epsilon Eridani b, orbits at 3.36 AU and is 0.88 times as massive as Jupiter. The problem with objects like these is their large orbital eccentricity. For example, the eccentricity of Epsilon Eridani b is 0.6 (compared with 0.048 for Jupiter). In other words, our Jupiter has an almost circular orbit, whereas Epsilon Eridani b orbits in an ellipse. In fact, the average eccentricity of the exoplanets discovered to date is 0.28 (with the eccentricities ranging from 0, for the hot Jupiters in perfectly circular orbits, through to 0.93, for a planet around the star HD80606). Compare this with the average eccentricity of planets in the Solar System: 0.08 (or 0.06 if we discount Pluto). Our Jupiter has a stable, nearly circular orbit — and it permits Earth to have a stable, nearly circular orbit too. If Jupiter were in a highly eccentric orbit, which seems the norm for a large-mass object orbiting more than about 0.2 AU from its star, then Earth might not exist.

Chapter 5

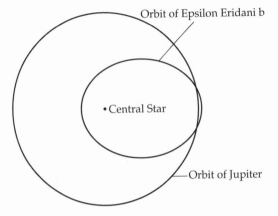

Orbit of Epsilon Eridani b

• Central Star

Orbit of Jupiter

FIGURE 47   *A comparison of the orbits of Jupiter and Epsilon Eridani b, drawn to the same scale. (Jupiter orbits the Sun with a semi-major axis of 5.2 AU; Epsilon Eridani b orbits its star with a semi-major axis of 3.36 AU.) Jupiter's orbital eccentricity is 0.048, although at this scale it appears to be circular. The orbital eccentricity of the planet orbiting Epsilon Eridani is 0.6 — which is noticeably elliptical.*

So had our Solar System contained either a "hot Jupiter" or an "eccentric Jupiter," the chances are high that Earth could not have sustained life for nearly 4 billion years. Earth's orbit would have been altered catastrophically. It is worth stressing, once again, that our observations are significantly biased. The Doppler techniques we use to discover other planetary systems are most effective precisely at finding (i) large-mass planets orbiting very close to the parent star and (ii) large-mass planets with highly elliptical orbits. Those objects provide the largest effects for our Doppler techniques to work on. A Jupiter-mass planet in a circular orbit at 5 AU from the parent star will — for the moment — be undetectable. We cannot yet deduce from these statistics that "good Jupiters" are rare. On the other hand, it is *possible* we were lucky; we found ourselves with a "good Jupiter" — one possessing a stable, circular orbit. Perhaps most planetary systems are not so lucky; perhaps "bad" Jupiters are the norm?

What about planetary systems with no Jupiter — neither good nor bad — at all? It is not clear whether planetary systems can form without also forming massive gas giants like Jupiter. Even if such systems *can* form, they may be no more conducive to life than are systems containing a "bad Jupiter." Our Jupiter has played two roles vital to life on Earth: that of deflector and that of water provider.

In its first role, Jupiter's great mass causes stray objects in elliptical orbits, which might otherwise hit Earth, either to be ejected from the Solar System or to have their orbits made circular and therefore less dangerous. And if neither of these things happens, Jupiter itself is the biggest target for rogue objects. In 1994, for example, comet Shoemaker–Levy 9 hit Jupiter; had it hit Earth, life on our planet would now be rather different.

In its second role, which it fulfilled early in the history of the Solar System, Jupiter caused asteroids to accrete into Mars-sized planetary embryos

FIGURE 48   In 1994, comet Shoemaker–Levy 9 hit Jupiter. Had it hit Earth, the devastation would have been immense.

with unstable elliptical orbits. Solar System objects in elliptical orbits are more likely to collide with objects in circular orbits; and some of the proto-planets collided with Earth. If such collisions were to happen now, the results would be cataclysmic. Back then, though, the results proved ulti-mately beneficial. The Moon may have been the result of one such collision, and our oceans may have been the result of other collisions. If recent work suggesting Earth's oceans came from asteroids is correct, then it implies that, without a Jupiter at the right distance to toss water-bearing asteroids our way, Earth might now be a desert.[196]

Computer simulations indicate that a Jupiter-mass planet forming in the very distant regions of a planetary system allows an Earth-mass planet to form with plenty of water — but only at 4 or 5 AU, which is far outside the habitable zone. So it seems that a planetary system needs not only a "good Jupiter," but one at exactly the right distance, otherwise the system's water is either trapped in an asteroid belt or is frozen on terrestrial planets. And as far as we know, if a planet has no liquid water, then it has no life.

*   *   *

So does the existence of Jupiter, our "Big Brother," explain the Fermi para-dox? As an explanation on its own, I doubt it — though of course it may yet be another factor causing life to be rare. My guess is that, as more data ar-rive, we will discover many planetary systems with "good Jupiters." And even if "good Jupiters" *are* rare, surely it is stretching matters to go from saying Jupiter played a beneficial role in the development of the Solar Sys-tem to saying a Jupiter-sized planet at about 5 AU is *essential* for life to exist on a terrestrial planet. Perhaps other arrangements of objects in a planetary system may lead to habitable zones. Our failure to discover these arrange-ments may simply be a failure of our imagination.

On the other hand, we see several pleasant coincidences in our Solar System — and Jupiter plays a part in most of them. Perhaps we have to thank Jove for a lot of things! The next section describes another reason why advanced life on Earth might not have developed without Jupiter.

Chapter 5

Solution 38: Earth Has an Optimal "Pump of Evolution"

When resonance occurs, a small input force can produce large deflections in a system.
Report on the collapse of the Tacoma Narrows Bridge

Jupiter plays a key role in another proposed resolution of the Fermi paradox — one that elaborates on an idea mentioned in the previous section. The suggestion is due to the physicist John Cramer.[197]

Large meteors sometimes hit Earth; but where do they come from? One idea is that they fall toward Earth from the Asteroid Belt — but for this idea to work, large numbers of asteroids must be perturbed from their stable orbits and then fall toward the inner part of the Solar System. Why should asteroids be pushed away from their stable orbits? No mechanism was known that could do this; then, in 1985, George Wetherill highlighted the importance of the gap in the Asteroid Belt at a distance of 2.5 AU.[198]

The rings of Saturn and the *Kirkwood gaps* in the Asteroid Belt were already well known. The gaps occur because of resonance effects. In the case of the gap at 2.5 AU, the resonance occurs because any asteroid at that distance orbits in precisely $\frac{1}{3}$ of the time Jupiter takes to orbit the Sun. Therefore, every third occasion a 2.5-AU asteroid reaches a particular position, Jupiter is in the same relative position. The gravitational nudge that Jupiter gives the asteroid is always in the same direction, and the effect is cumulative. It is like pushing a swing at precisely the right frequency: the effects build up, and the amplitude of the swing increases. Over time, therefore, the orbit of an asteroid at 2.5 AU becomes unstable, and it moves away — and the Asteroid Belt is eventually cleared of objects in this region. (Any asteroid wandering into this region from elsewhere is eventually ejected by the same mechanism.) The Kirkwood gap at 2.5 AU is due to a 3:1 resonance; other gaps, based on other resonances with Jupiter, also exist.

Where do the asteroids go when they are ejected from the Kirkwood gap at 2.5 AU? Calculations show there is a high probability of their orbits crossing the orbit of Earth. In other words, there is a chance that these asteroids hit Earth — with catastrophic consequences.

However, although the effects of an asteroid impact can be disastrous for any creatures that happen to be around, in the *long* run the impacts may be beneficial. After all, if the meteor impact of 65 million years ago had not happened, then Earth might still be home to dinosaurs, and mammals might still be scraping a living at the margins of a lizard-dominated world. Cramer points out that there may be geological periods when nothing much happens to species; evolution appears to take the commonsense attitude of "if it ain't broke, don't fix it." It is primarily at crises points, when for some reason the environment changes, that evolution works quickly and new species arise to take advantage of altered conditions. Evolution, in Cramer's words, seems to be "pumped" by cycles of crises and stability. And, he suggests, an ideal pump is one that drives evolution through major crises every 20 to 30 million years. Asteroids from the 3:1 Kirkwood gap may provide a pump at exactly the right rate.

FIGURE 49   A montage of images of Eros; the images were taken over three weeks as the NEAR spacecraft approached the asteroid. Near-Earth asteroids like Eros are relatively few in number. Most asteroids are in the "main belt," orbiting the Sun in a torus between Mars and Jupiter. It is these "belt" asteroids that can be perturbed from their orbits by the gravitational influence of Jupiter — with potentially devastating results.

If Cramer's idea is correct — and he would be the first to admit that the idea is speculative — it constitutes another reason why life on Earth might be special. Not only might life require an Earth-like environment, the environment might need to occur in a system with planetary masses and orbits that produce a resonance in an Asteroid Belt at just the right rate. If the "pump of evolution" runs too fast — and asteroids hit a life-bearing planet too often — then life never has a chance to evolve intelligence. If the pump runs too slow — and asteroids hit a life-bearing planet too rarely — then life becomes stuck in a rut. The result is a planet full of trilobites or cockroaches or dinosaurs (or, more likely, creatures differing from terrestrial creatures in a myriad of fascinating ways). As long as these creatures were successful, in an unchanging environment there would be no "need" for them to adopt new modes of behavior, and no "need" for them to develop intelligence and thence radio telescopes or starships.

The existence of the Asteroid Belt is due to Jupiter: the Belt is the remnants of a protoplanet whose formation was aborted because of Jupiter's own formation. And the 3:1 resonance in the Belt is also due to Jupiter. If there is such a thing as a "pump of evolution," and if it is tuned to the right level in our planetary system, then we have Jupiter to thank for it.

SOLUTION 39: THE GALAXY IS A DANGEROUS PLACE

I am become death, the destroyer of worlds.
Bhagavadgita

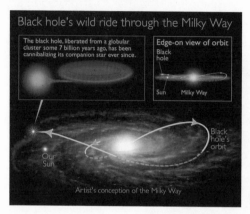

Black hole's wild ride through the Milky Way

The black hole, liberated from a globular cluster some 7 billion years ago, has been cannibalizing its companion star ever since.

Edge-on view of orbit

Black hole

Sun   Milky Way

Black hole's orbit

Our Sun

Artist's conception of the Milky Way

FIGURE 50   *Black holes may be lurking out there in interstellar space.*

A key realization of modern astronomy is that the Universe is a perilous place. We now know that violent phenomena are common and pose a variety of threats. A stray black hole wandering into a planetary system would devour the planets and any life they harbored. (We know black holes exist. Some astronomers estimate that a million of them may be wandering through interstellar space. Could one of them be heading our way?) Neutron stars called magnetars would pose an interesting threat if they came too close. (On 27 August 1998, several orbiting detectors recorded radiation from the magnetar SGR 1900+14. The radiation came within 30 miles of Earth's surface. Fortunately, our atmosphere shielded us, as it does from a variety of forms of cosmic radiation. SGR 1900+14 is 20,000 light years away, so would our atmosphere have saved us had the magnetar been closer?[199]) A galaxy might possess a violently active nucleus, which is quite deadly. (The central region of our own Galaxy, although not as active as objects like blazars, for example, is nevertheless inhospitable. Close to the center, stars are so crowded that the night sky would be bright enough to read by; closer still, and you meet the accretion disk of a million-solar-mass black hole. This is why the inner edge of the GHZ is defined by the point where the violent central regions are no longer a threat.)

Could this be the explanation of the Fermi paradox? Might the random violence of an uncaring Universe explain the silence? Are civilizations destroyed before they can reach us?

The three mechanisms mentioned above — stray black holes, magnetars, and active galactic nuclei — do not by themselves, or as a group, explain why our Galaxy is silent. Black holes and magnetars might pose a threat to individual stars or stellar groups over the course of the Galaxy's lifetime, but they cannot act as a Galaxy-wide sterilizing agent; and while the center of the Galaxy is probably a place to avoid, it seems to provide no threat to life way out here in the spiral arms, some 30,000 light years away

3 Arc Minutes

FIGURE 51 A
Hubble Space
Telescope image of the
center of the galaxy
NGC 253. The central
region of this galaxy
is violently energetic,
and is not likely to be
a hospitable place for
life.

from the action. On the other hand, two other mechanisms — supernovae
and gamma-ray bursters — might resolve the Fermi paradox.

Supernovae

A supernova is the cataclysmic explosion of an aging star. Such explosions
are powerful and occur rather frequently on an astronomical timescale: the
Galaxy on average is host to one or two supernovae per century.

There are two types of supernova. A Type Ia supernova results when
a white dwarf in a binary system reaches a critical mass after sucking material from its companion. A violent thermonuclear explosion ignites and
blows the star to bits. A Type II supernova occurs in the later stages of the
life of massive stars. When the core of a massive star no longer produces
enough energy to support itself against the relentless force of gravity, the
star collapses under its own weight. The core forms a dense neutron star or
even a black hole; the outer layers of the star rebound from the core at high
velocity and head off into space, where they become part of the interstellar
medium. (Life on Earth would not exist were it not for an ancient Type II
supernova that seeded space with heavy elements cooked up in its core.)
The details of the two types of explosion are different, but both types radiate large amounts of energy. Over the course of a few weeks, a supernova
can release as much as 10^{44} J in a variety of forms.

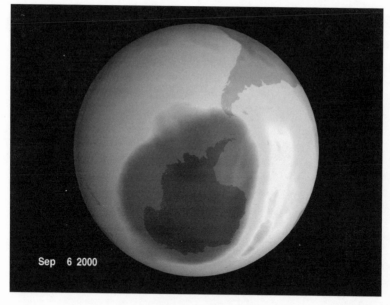

FIGURE 52
Ozone
depletion
over the
South Pole
in 2000. A
nearby
supernova
could
reduce
ozone levels
over the
entire globe.

A nearby supernova might be disastrous for life on Earth. One estimate is that any supernova exploding anywhere within about 30 light years of Earth could destroy most surface life on our planet.

However, the mechanism of destruction is not obvious. For example, although a Type Ia supernova is intrinsically the brightest type of supernova, even at maximum brightness it would have to be no farther away than one light year to appear as bright as the Sun. On an astronomical scale this is extremely close, so we have nothing to fear from supernova optical photons. Type II supernovae emit vast numbers of neutrinos, and perhaps the large neutrino flux from a nearby supernova could have deleterious effects upon organisms. But it is difficult to believe that neutrino fluxes can lead to mass-extinction events. No, the real threat is the enormous amount of gamma-radiation that a nearby supernova would dump into Earth's atmosphere. Direct gamma-radiation from the explosion would probably not harm us, because the upper atmosphere provides an effective shield. However, the gamma-rays would cause atmospheric nitrogen to dissociate, the nitrogen would then react with oxygen to form nitric oxide, and the nitric oxide would react with ozone — thus rapidly depleting the ozone layer. Ozone levels could be reduced by as much as 95% for several years. With Earth's ozone layer down, surface life would have nothing to protect it from lethal UV rays from the Sun. The supernova, in other words, kills by a classic one–two punch: first the gamma-radiation lowers our defenses, then the solar UV radiation devastates multicellular life.

As we will discuss later, there have been several mass-extinction events since multicellular life took to the land. Can any of these be blamed upon the effects of a local supernova? It is difficult to say with certainty. As seems increasingly probable, the last mass extinction — the one in which dinosaurs perished — was due to the effects of a meteor impact. Perhaps the other great die-offs were caused by similar impacts; or perhaps they were due to climate change; or perhaps they were just chaotic events that can happen in complex systems. There is no evidence linking mass extinctions to the after-effects of supernovae. Even if supernovae *can* cause mass extinctions, it is not certain whether the extinctions pose a long-term threat to the emergence of intelligence. Perhaps, indeed, supernovae are *necessary* for intelligent life. Maybe, to use Cramer's phrase, they constitute another "pump of evolution." For the moment, though, let us assume that a nearby supernova can cause a mass-extinction event, and that such an event slows the development of intelligent life.

Since all stars, including the Sun, move through space, over the course of aeons random stellar motions will bring the Sun close to a supernova. Eventually, a supernova *will* explode close to Earth. (In case any readers are worried, no star presently within 60 light years of us will go supernova within the next few million years.) The critical question is: how often is a supernova event likely to occur close enough to Earth to cause a mass-extinction event? Typical estimates are that a supernova event will occur within 30 light years of Earth on average every couple of 100 million years. If that is true, we have another question to ask. Why are *we* here?

One answer to this question could be simply that the calculations of the frequency of supernovae are wrong; or (which is quite likely) perhaps we do not fully understand the effects of a nearby supernova. In this case, there is no implication for the Fermi paradox. But perhaps we are here because Earth has been extremely lucky; perhaps Earth has not seen a really close supernova since the emergence of life on land. If this is true, then we can resolve the Fermi paradox by saying that *every* other life-bearing planet has been less fortunate than Earth.

However, resorting to luck is a poor sort of explanation. And there is no astrophysical evidence to suppose Earth has been particularly fortunate with regard to supernovae. If *we* have been lucky, then there is no reason to suppose that, in the past, other regions of the Galaxy did not also have a run of good luck. Indeed, if we accept that intelligent life is common, then supernovae are just not effective enough to explain the Fermi paradox. Inevitably, by the blind workings of chance, some civilizations will never come close to a supernova and will thus have the time to develop space travel. And once they colonize other parts of the Galaxy, no supernova can stop them. (Hence, the threat from supernovae is another motivating factor for ETCs to engage in interstellar colonization! Once a civilization has colonized stars within a radius of about 30 light years of the home world, they will survive the effects of a local supernova.)

What we need if we want to explain the Fermi paradox is a mechanism that can affect life on *every* planet in the Galaxy, without exception. If there were some mechanism that generated a sufficiently powerful Galaxy-wide sterilizing event it could operate fairly infrequently (every few hundred million years, say) and remain an explanation for the Fermi paradox. Multicellular life would be eradicated before intelligence had the chance to arise; a civilization could never advance to the stage where it might develop effective countermeasures to the threat. Putative ETCs would not have had billions of years to colonize the Galaxy; instead, they would have the few hundred million years since the last sterilizing event. In essence, the "Universal Clock" would be reset every time a sterilizing event took place.

It seems unbelievable that any phenomenon could cause such widespread devastation. Unfortunately, astronomers now know of a potential Galaxy-wide sterilizing mechanism: the devastating power of a gamma-ray burster (GRB).

Gamma-Ray Bursters

Gamma-ray bursters were discovered by accident more than 30 years ago, but until recently their origin was completely unknown.[200] Even now, the precise physical origin of GRBs is a matter for intense debate among astronomers. Whatever the progenitor event may be, the important fact about a GRB is this: the GRB fireball is the most powerful phenomenon in the known Universe. A GRB pours out more energy in a few seconds than the Sun will generate in its entire lifetime. A GRB shines so brightly that our detectors can see them from halfway across the Universe. All the GRBs we have detected so far seem to have occurred in distant galaxies; if one occurred in our Galaxy, it would be bad news. We need to ask two questions. First, how frequently do GRBs occur in our Galaxy? Second, if our Galaxy hosted a GRB event, just how bad would things be?

Calculating the frequency of occurrence of Galactic GRBs is a typical Fermi problem! It happens that a galaxy hosts a GRB about once every 100 million years. Interestingly, this rough timescale is pretty much the timescale between mass-extinction events on Earth. People have suggested, therefore, that GRBs might be responsible for mass extinctions.

The Frequency of Gamma-Ray Bursters

A gamma-ray detector such as BATSE (Burst and Transient Source Experiment) on board NASA's orbiting Compton Gamma Ray Observatory detects on average one GRB per day. BATSE covers about one third of the sky, and therefore about three GRBs occur in the Universe each day — or about 1000 each year. As a rough estimate, we can suppose that there are 10^{11} galaxies in the Universe; so on average there are 10^{-8} GRBs per galaxy per year. In other words, to a first approximation with which Fermi would be happy, a typical galaxy will host a GRB about once every 100 million years. (This calculation assumes that GRBs emit their energy equally in all directions. If GRBs emit their energy in a beam, as some theories suggest, then the GRBs we detect would be those with beams that happen to be pointing toward us. The total GRB-event rate would therefore be much higher, to account for those GRBs with beams pointing in some other direction. For our purposes, though, we do not need to consider this point.)

The awesome power released by GRBs means that, even if one occurred at a large distance from Earth, our planet would still be bathed in radiation. Moreover, the same GRB could cause devastation throughout the Galaxy. The pessimists suggest that a GRB could sterilize the Galaxy.

Nevertheless, this suggestion is very much open to debate. Gamma-ray bursters are undeniably more powerful than supernovae, so they could be at much greater distances and still inflict the same sort of damage to the ozone layer, through the same processes. But there is a difference.

FIGURE 53 *Did a gamma-ray burster kill the dinosaurs? In this artist's impression, T. Rex looks up at the brief flash of a burster. A much more probable scenario, however, is that a meteor impact caused the mass extinction at the end of the Cretaceous period. Whether gamma-ray bursters (or supernovae) caused earlier mass extinctions is not known.*

While a supernova event occurs over a rather long period of time, a GRB pumps out most of its energy in less than a minute. Therefore only half of a planet will be affected directly by a burster; the other half is safe from the blast, as it is shielded by the mass of the planet. Of course, the damage from the affected side of the planet *might* propagate and cause worldwide destruction; and secondary effects *might* cause further problems. But with our present state of knowledge, it is just as easy to argue that a planet's ozone layer would protect surface life from the effects of a GRB — unless the GRB occurs *too* close, of course, in which case the planet is toast.

Suppose we accept that a GRB can destroy all higher life-forms throughout a galaxy. Combine this with the prediction from some theories of GRB formation that bursters were more frequent in the past, and you have the resolution of the Fermi paradox proposed by James Annis.[201] The proposal is simple: In the past, GRBs effectively sterilized planets before any life-forms in a galaxy had the chance to develop intelligence. Only now that the event rate has decreased, and GRBs are less common, has there been time for technologically advanced civilizations to arise. With Annis' proposal, there is nothing necessarily special about Earth or humanity; there may be tens of thousands of ETCs in our Galaxy at or near the same stage of development. All of them will have had the same time as life on Earth to develop: the amount of time since the last GRB exploded in the Galaxy.

Personally, I think it unlikely that GRBs are capable of sterilizing whole galaxies, and thus I do not accept that GRBs by themselves resolve the Fermi paradox. It is undeniable, however, that GRBs occur and are astonishingly powerful; they would certainly sterilize any planet unlucky enough to be nearby. The SETI optimists — those who argue that intelligent, technologically advanced civilizations are common — thus have to face an unpalatable conclusion: over the course of a Universal Year, many of those civilizations must have been within distance of a GRB. Countless numbers of advanced civilizations must have been consumed by fire.[202]

Solution 40: A Planetary System Is a Dangerous Place

Man is never watchful enough against dangers that threaten him every hour.
Horace, *Carmina*, II.13

Destruction may come not just from the distressingly long list of celestial hazards. Some threats are much closer to home. We have already mentioned the most obvious worry: meteorite impact. Tiny meteorites fall to Earth every day; medium-sized objects land every few years; large objects — say, 20 km wide — hit Earth every few hundred million years. Although large meteorites only hit Earth infrequently, when they *do* hit they cause total devastation. If a 20-km-wide asteroid hit Earth today, it would almost certainly kill every human being. Multiply the small chance of an event occurring by the number of people it would kill, and you arrive at the probability of death per person for the event. It turns out that, averaged over a human lifetime, the chance of being killed by meteorite impact is about the same as dying in an aircraft crash. Paradoxically, we spend vast amounts of money on air safety, yet essentially nothing on detecting the near-Earth objects that could destroy our civilization.

FIGURE 54   *If a meteor like this one hit Earth, human life would almost certainly be wiped out.*

Presumably, ETCs also have to contend with the threat posed by meteorite impact, as these objects are probably common in planetary systems. But there are many other hazards, and below I discuss a few more.

Snowball Earth

The threats need not even come from space. Recent evidence — particularly the discovery of glacial debris near sea level in the tropics — suggests that, over geological history, Earth has repeatedly been covered in a layer of ice. One event may have happened 2.5 billion years ago, and there may have been four of these *Snowball Earth* events in the last 800 million years, with each episode lasting for 10 million years or more. Do not mistake these events with the textbook images of the last Ice Age; compared to a

Snowball Earth, the last Ice Age was positively tropical. During a Snowball Earth, a kilometer-thick layer of ice covers the oceans, and ice even covers equatorial oceans (though perhaps not to the same depth). Average temperatures drop to $-50°$ C. Most organisms are unable to cope with such conditions, and life can hang on only by the thinnest of threads — perhaps around volcanoes, or under clear thin ice at the equators.[203]

FIGURE 55
Melting ice floes in open water. On a Snowball Earth conditions at the equator would, at best, be like this. All else would be covered in thick ice.

How our planet can descend into a Snowball Earth is well understood: The ice cover can increase for a variety of reasons, and when it increases the ice reflects an increasing amount of sunlight straight back out into space. This decrease in solar heating of the surface causes the temperature to drop and more ice to form. Once a critical amount of ice cover is reached, a "runaway icehouse" effect takes place and the planet descends into a Snowball Earth event. What is difficult to understand, and what caused scientists to dismiss the idea of a Snowball Earth for many years, is how the planet can *escape* from the ice cover. Once Earth is encased in ice, most of the sunlight falling on the planet is reflected into space before it can warm the surface. The solution came with the realization that volcanic activity does not stop during a Snowball Earth event. Volcanoes pump out vast amounts of carbon dioxide — a greenhouse gas. Of course, volcanoes are still belching out carbon dioxide, but under normal conditions this CO_2 is absorbed by falling rainwater, which eventually carries it to the ocean where it becomes locked up in solid carbonate deposits on the ocean floor. On a Snowball Earth there is no liquid water to evaporate, and therefore no clouds, and therefore no rain: for 10 million years, maybe more, the CO_2 from volcanoes would build up in the atmosphere. Eventually, there would be about a thousand times more atmospheric CO_2 than in today's atmosphere. The temperatures would rise and quickly melt the ice: from icehouse to greenhouse in a geological instant.

The implications of the Snowball Earth hypothesis are profound, and we shall examine some of them later.

Super-volcanoes

Although volcanoes proved to be life's savior during the Snowball Earth events of the Neoproterozoic era, more recently they almost proved disastrous for intelligent life on Earth: they have almost wiped out *Homo sapiens*.

Recent research indicates that, genetically, humans are all remarkably similar. To explain this lack of genetic diversity, some biologists have suggested that *Homo sapiens* must have emerged from a "genetic bottleneck" about 75,000 years ago. A bottleneck occurs when the size of a population reduces dramatically. In the case of our species, the total number of humans alive on Earth may have dropped as low as a few thousand. We almost became extinct.

If this bottleneck really did occur, then we do not have to look far for a smoking gun that may have caused it. The Toba volcano in Sumatra erupted 74,000 years ago; so great was the eruption, it earns the title of a "super-volcano." The eruption was *much* more violent than recent volcanoes like Mount Pinatubo and Mount St. Helens. Climatologists have suggested that a super-volcanic eruption can cause a volcanic winter — similar in effect to a nuclear winter, but without the radiation. It is not implausible that the years of drought and famine following such an explosion could drive a pre-technological human species to the brink of extinction.

Mass Extinctions

Meteor impact, global glaciation, super-volcanoes. Even on a placid planet like Earth, life has to contend with a lot. Sometimes, whether the cause is one of the three mechanisms mentioned above, or one of the celestial agents of destruction, life barely hangs on.

Life on Earth has suffered several mass extinctions — a mass-extinction event being defined as a period that sees a significant reduction in Earth's biodiversity. There have been fifteen such events over the past 540 million years. (There may have been many more extinctions earlier in Earth's history, particularly in Snowball Earth events, but only in the past half billion years have creatures with hard skeletons become common; so only relatively recently could creatures become fossils. Indeed, the time since the Cambrian age is known as the Phanerozoic era, from Greek words meaning "visible life." The 4 billion years before the Cambrian age is known as the Cryptozoic era, from Greek words meaning "hidden life." For most of Earth's history, virtually all organisms lived and died without leaving traces.) In six great mass-extinction events, more than half of all species then alive were killed.[204] These six events are, in chronological order, the Cambrian, Ordovician, Devonian, Permian, Triassic and Cretaceous.

The Cambrian extinction (actually two extinctions) occurred 540 to 500 million years ago. Their precise cause is uncertain, but in some ways they were the most serious of the mass extinctions. During the *Cambrian explosion*, a time of immense biological innovation, Nature experimented with many different body plans; perhaps as many as a hundred different animal phyla evolved. All the animal phyla we are familiar with today emerged during the Cambrian explosion, and no new phyla have evolved since. But during the Cambrian extinctions some of these phyla — each containing species that seem bizarre and even nightmarish to our eyes — died out.[205]

The Ordovician extinction 440 million years ago and the Devonian extinction 370 million years ago both saw more than a fifth of the marine families disappear. The effects on land life are less well known, mainly because the fossil record is so poor for these ages. Nor is the cause of these extinction events known; if impact events caused them, no trace of the resultant craters has been found.

The Permian extinction 250 million years ago was even more severe than the Cambrian extinction. Perhaps more than 90% of marine species became extinct; eight of the 27 orders of insects were lost; the loss was devastating. The cause of this catastrophic event is uncertain; several mechanisms, possibly acting in synergy, have been proposed to explain this global catastrophe.

The Triassic extinction 220 million years ago saw significant reductions in the number of marine and land species. Many scientists believe a meteorite was the cause of this extinction event.

The Cretaceous extinction 65 million years ago is the most celebrated and most well-known of all the mass extinctions. This event saw the end of the age of the dinosaurs (and provided the conditions that led to the rise of the mammals). Almost certainly, the cause of this extinction was the after-effects of a large meteorite impact. There are several reasons for believing in the impact theory of this extinction event. First, the 200-km-wide Chicxulub crater on the Yucatán peninsula in Mexico is of precisely the right age. Second, no matter from where in the world they are drawn, rock samples from the Cretaceous–Tertiary boundary show a high concentration of iridium, which is what one would expect if a large asteroid hit Earth. Third, many of the same sites contain shocked quartz grains — another sign of violent impact. Fourth, geologists often find fine soot particles in clays from the Cretaceous–Tertiary boundary — particles that could have come only from burning vegetation; the implication is that much of Earth's plant matter was on fire.[206] The immediate aftermath of the impact would clearly have killed large numbers of organisms. The precise mechanism for eradicating large numbers of species is less clear; it could have been atmospheric change, a nuclear winter, large-scale long-term fires, acid rain, a combination of these effects ... or something else entirely. The effects were also dependent upon when and where the meteorite struck Earth, and also on the mass and velocity of the meteorite. Had the meteorite struck just a few hours later, the effects might have been less deadly; had the meteorite been just twice as large, the extinction of life might have been total.

Extinctions and the Fermi Paradox

It is difficult to say what we can learn from these extinction events. They seem to be different in character, cause and severity. Only in the cases of the Cretaceous and Permian events are there obvious causal mechanisms for the extinctions. The other extinctions may have been caused by something quite different; after all, we have considered many potential threats. Life-forms on other planets presumably face the same hazards, and they may face risks that life on Earth has been spared. For example, some planetary systems may have life-bearing planets in orbits that become chaotic — and a mass extinction would be probable. Or a change in the rotational rate of a planet might trigger a mass extinction. Anything that causes extensive climate change — either a global cooling or warming outside of temperatures that are tolerable for animal life — might induce a mass extinction. Perhaps the lesson is simply that planetary systems are dangerous: over the course of billions of years, mass extinctions are *inevitable*.

It is a short step from arguing that mass extinctions are inevitable to arguing that they play a role in resolving the Fermi paradox. In fact, people have used the idea of mass extinctions to suggest two quite antithetical solutions to the paradox. The straightforward suggestion is that mass-extinction events have impeded the development of intelligent life on other planets. The more subtle suggestion is that, in the immortal capitalization of Sellars and Yeatman, mass extinctions are a Good Thing that occur too infrequently on other planets! (At least, the right *sort* of extinction events happen too infrequently.)

It is easy to understand why mass extinctions might be a Bad Thing. Many people would argue that life — at least life as we know it — has only two defenses against mass extinction. The first defense is simplicity: this is the approach taken by prokaryotes (see page 190), which have survived for billions of years. Bacteria have essentially kept their single-cell body plan over the aeons; indeed, it is probable, though difficult to prove conclusively, that modern bacteria are genetically identical to the earliest living cells of 3.7 billion years ago. Their ability to evolve biochemical responses to new environmental challenges enables them to take most things Nature can throw at them. Only a catastrophe on a massive scale would remove all prokaryote life from Earth. On the other hand, we cannot communicate with bacteria. When considering Fermi's question, we are interested in *complex* multicellular life-forms. How do *they* survive the slings and arrows of a billion years of fortune?

The second defense against mass extinction is diversity — an approach taken by animals and plants. If a phylum contains many different species, if it has different ways of earning a living, then there is a chance that one or two of the species will survive the extinction event. Later, the diversity of the phylum can be replenished. So even though animal and plant life is less hardy than bacterial life, and is much more susceptible to extinction, in the long run it can survive. (Perhaps the main reason the Cambrian extinction was so deadly is because, although there were many different phyla, each phylum contained only a few species. Entire phyla became extinct because they contained insufficient diversity. It is something of a theme of this book: do not put all your eggs in one basket.)

We have no idea how evolution has proceeded on other planets; but perhaps Earth is rare in having phyla with many different species. (See page 181 for a reason why this might be the case.) Complex life on other worlds may be less likely to survive the inevitable extinction events. We can imagine worlds that are home to many different, weird-looking, truly *alien* creatures — creatures possessing a variety of peculiar body plans. There might be a large number of phyla on such worlds — phyla that took aeons to evolve to their present state. But if those phyla are represented by only a few species — well, when the meteorite strikes, or the climate heats up, or the planet's obliquity changes, those phyla may well die out. Maybe Earth has just been lucky (there is that word "lucky" again). This is a gloomy resolution of the Fermi paradox.

We have encountered the more subtle suggestion regarding mass extinctions — namely, that they may be *necessary* for the development of intelligent life – when we discussed the suggestion of a "pump of evolution." Of course, it would be no fun being around when a 20-km-wide asteroid smashes into Earth or global temperatures plummet. But in the long run — a run measured in tens of millions of years — life might benefit from such catastrophes. After the deluge, new and radically different forms have a chance to evolve; Nature can use the changed environment to create and experiment with different species, and perhaps even different body plans. Certainly, following mass-extinction events, biodiversity has always regained the pre-extinction level and then exceeded it.

One currently controversial suggestion is that two key events in the history of life on Earth — the development of the eukaryotic cell, and the Cambrian explosion (more of which in later sections) — were a direct result of the escape from Snowball Earth events. The chemical changes a Snowball Earth would cause in the oceans, the genetic isolation of species, the great environmental pressure on life, the rise in temperature and the rapid melting of ice — all these factors might combine to produce a time of rapid evolutionary activity. According to some scientists, neither animals nor higher plants would exist today if it were not for past Snowball Earth events.

Perhaps the "right" global glaciation events are uncommon on other planets. A planet has to be in the CHZ, it has to have oceans of water, it has to descend into an icehouse, and it must possess active volcanoes spewing out greenhouse gases to remove the ice. Perhaps the norm for most water-planets is a descent into a Snowball with no means of escape; the mass extinctions would be total.

Chapter 5

The Holocene Extinction

It is impossible to discuss past mass-extinction events on Earth without mentioning the *Holocene extinction*. The Holocene epoch encompasses the last 10,000 years, up to the present day. In other words, we are living through yet another mass-extinction event. In this case the cause is clear: human activity. We hunt species to extinction; we introduce alien species into new ecosystems and cause havoc; and, most importantly, we destroy habitats. It does not *feel* as if we are in the midst of a mass extinction because, on an individual scale, 10,000 years is a long time. On a geological scale, though, it is an instant. Under some estimates, the rate at which species are becoming extinct is now 120,000 times the "normal" or "background" rate.[207] Many of the species made extinct by our destruction of rain forests have never even been documented. If the current rate of extinction is maintained, and the destruction of the rain forests continues, then global atmospheric and climatic effects seem certain to occur. It is then highly probable that *Homo sapiens* will be one of the species that joins the extinction. Harking back to a previous solution discussed in the book, perhaps a general evolutionary law is that intelligence extinguishes itself.

SOLUTION 41: EARTH'S SYSTEM OF PLATE TECTONICS IS UNIQUE

What we want is a story that starts with an earthquake and builds to a climax.
Samuel Goldwyn

Our planet has been destructive in recent years. Earthquakes in Turkey and India have caused huge loss of life; smaller quakes in America and Japan have caused inconvenience; and as I write, Mount Etna is spewing forth lava that threatens the livelihood of several hundred villagers.[208] It therefore seems strange that some geologists consider the existence of *plate tectonics* — the process that gives rise to earthquakes and volcanic eruptions — to be necessary for the existence of complex life. But there is a serious reason for believing that three phenomena — life, water oceans, and plate tectonics — are linked. And this linkage may be unique to Earth.

*   *   *

The various planets of the Solar System have different methods of disposing of their internal heat. In Earth's case, the heat generated by radioactive decay in the interior is transported by the convective method known as

180

plate tectonics. Consider what happens near a mid-ocean ridge. Hot material from the deep mantle region of Earth is brought to the surface in a convection cell, and at the surface it spreads out and solidifies into ocean crust — it becomes part of the lithosphere. Over geological timescales, the new material floats on the hot mantle underneath it and moves away from where it was born. During this process it cools and collects masses of igneous rocks. The material becomes heavier, and after many tens of millions of years it sinks back, under its own weight, deep into the mantle at places called *subduction zones*. Eventually, the cycle repeats. On geological timescales, the outer regions of our planet resemble one of those kitsch lava lamps.[209]

Some scientists think plate tectonics may be the most important requirement for the development of animal life. There are several reasons why plate tectonics might be vital. Let us look at just three of them.

First, the mechanism of plate tectonics seems important in the creation of Earth's magnetic field. The theory of planetary magnetism is formidably complicated, but, in essence, planets generate a magnetic field by means of an internal dynamo. Such a dynamo requires three things: the planet must rotate, it must contain a region with an electrically conducting fluid, and it must maintain convection within the conducting fluid region. It is difficult to be sure, but in Earth's case it seems likely that without plate tectonics the convective cells would cease to export heat to the surface, the dynamo would not function, and Earth's magnetic field would be a tiny fraction of its present value. The relevance of all this is simple: Earth's magnetic field helps prevent high-energy particles in the solar wind from scattering atmospheric particles into space; over time, such sputtering could cause the Earth's atmosphere to dissipate. In short, without Earth's magnetic field surface life might not have evolved.

Second, plate tectonics, or *continental drift*, created Earth's continents — and continues to refresh them. Continents are important. A world with a mixture of oceans, islands and large continents is more likely to offer evolutionary challenges than is a world dominated solely by water or land. Furthermore, plate tectonics causes environmental conditions to alter, and thus helps promote speciation. (For example, suppose the splitting of a piece of land from a continental land-mass results in a particular species of bird living on both the new island and the original continent. Over time, the environment on the island will differ from the continental environment; the birds will face different challenges and will evolve in different ways. Over time, there will be two species where before there was one.) Plate tectonics thus promotes biodiversity, which, as we have seen, is important during times of mass extinction. The larger the number of species, the greater the chance of *some* of them surviving the extinction event.

Third, and most important, for a billion years or more plate tectonics has played a key role in regulating Earth's surface temperature. The climate on our planet has long been balanced on a razor's edge. If the temperature drops too much, and the icecaps begin to increase in size, then a runaway icehouse effect may occur: Earth freezes. If the temperature increases too much, and the oceans start to simmer, then the extra water vapor in the atmosphere causes a runaway greenhouse effect: Earth boils. Certain prokaryotes might survive these temperature extremes, but complex life-forms can flourish only over a much narrower range of temperatures. Plate tectonics, some scientists argue, has a fine-tuning mechanism that keeps the planetary thermostat set "just right" for animal life.

FIGURE 56   *Mount Etna, in Sicily, is Europe's largest volcano. Although volcanoes like this can be tremendously destructive (though nowhere near as destructive as super-volcanoes), the underlying mechanism of plate tectonics that gives rise to them may be vital to life.*

The way plate tectonics controls temperature is rather complicated, and more than one mechanism is involved.[210] The key role it plays, however, is in its regulation of atmospheric carbon dioxide. CO_2 is an effective greenhouse gas: if the atmosphere contains too much CO_2, then global temperatures can rise (as mankind seems hell-bent on demonstrating experimentally). On the other hand, if there is too little atmospheric CO_2, then Earth fails to benefit from the greenhouse effect, and the planet cools.

Now, CO_2 does not remain in the atmosphere indefinitely. Carbon dioxide reacts with water to form carbonic acid; rainfall thus "washes" it out of the atmosphere. This carbonic acid weathers the rocks on Earth's surface, and the chemical products of this weathering get transported by rivers to the ocean. The products end up as calcium carbonate ($CaCO_3$) and quartz (SiO_2) on the seafloor, both through the formation of rocks and through the formation of the shells of living organisms. Eventually, the plate tectonics mechanism causes this $CaCO_3$ and SiO_2 to be subducted down into the depths of the Earth. Thus, atmospheric CO_2 is removed. But that is not the end of the story! The high temperatures and pressures deep within Earth convert the calcium carbonate back into CO_2 and CaO. Plate tectonics then recycles the CO_2 — and lots of other useful materials — by creating volcanoes (which vent tremendous amounts of CO_2; as we saw earlier, this mechanism allowed an escape from Snowball Earth episodes.)

If the atmospheric CO_2 were not replaced, Earth would undergo a global cooling. But what if too much CO_2 is put back into the atmosphere? Do we not run the risk of a runaway greenhouse effect? It turns out that, as the planet warms, the chemical weathering of rocks increases — which causes more CO_2 to be removed from the atmosphere, which causes the planet to cool (thus slowing the rate at which CO_2 is removed from the system, which causes the planet to warm ... and so on, in a classical feedback mechanism). This CO_2–silicate cycle is rather complicated, and the details are not fully understood, but the cycle seems to be crucially important for the long-term stabilization of global temperature.

One can argue that the development of animal life here on Earth *required* plate tectonics — to promote biodiversity, to generate a magnetic field, to stabilize global temperature, and so on. And yet there is nothing inevitable about plate tectonics. Only Earth, as far as we know, uses this mechanism to dispose of its internal heat. Perhaps the process is rare, and other planets lack animal life because they lack plate tectonics.

We do not know how frequently plate tectonics will occur because we lack a good general theory of the process. The type of questions one might ask — How does the existence of plate tectonics depend on a planet's mass? How does it depend on the chemical composition of the mantle? — cannot be answered with present models, so it is impossible to provide a good estimate of how many planets might develop, and maintain, plate tectonics. In the absence of hard facts, from either experiment or theory, one can argue either way. Some scientists believe the titanic collision that formed the Moon laid the seeds from which plate tectonics developed; in this case, plate tectonics may be rare. On the other hand, the basic conditions for plate tectonics seem relatively simple: a planet should have a thin crust floating on top of a hot, fluid region undergoing convection due to rising heat from the core. Perhaps water oceans are also necessary to "soften" the crust and allow subduction. Such conditions are probably not rare. Scarce, perhaps, but not rare. In other words, we simply do not know whether or not plate tectonics is common.

Even if plate tectonics *is* rare, does it necessarily follow that animal life is rare? Although plate tectonics seems to have played (and continues to play) a beneficial role for the development of life on Earth, is it the *only* mechanism that can provide these benefits? Plate tectonics is an extremely complicated and poorly understood process; the very existence of the CO_2–silicate cycle has only been known about for two decades. In cases like these, it often happens that there is more than one way to skin a cat. Maybe right now the scientists of a planet orbiting some anonymous M-class star are marveling at the cooling mechanism of their world and how it almost miraculously stabilizes their global environment.

My guess is that — like so many factors we have discussed — the possible rarity of plate tectonics is by itself insufficient to provide an answer to the Fermi paradox. But it may be yet another factor making it less probable that ETCs will develop on other planets.

SOLUTION 42: THE MOON IS UNIQUE

How like a queen comes forth the lonely Moon.
George Croly, *Diana*

The last time I checked, astronomers had found 68 satellites orbiting the planets of the Solar System. It therefore seems absurd to suggest that our Moon is unique, much less that the Moon has anything to do with the Fermi paradox. Yet for decades there has been a nagging suspicion that what makes Earth special is the Moon.

To understand why the existence of the Moon might resolve the Fermi paradox, we need to answer three questions. First, in what way is the Moon unusual? Second, how probable is it that satellites similar to Earth's exist in other planetary systems? Third, in what way might the existence of the Moon have been *necessary* for the development of intelligent life?

The Double Planet

Astronomy texts tell us that the Solar System contains nine planets, but this number flatters the outermost "planet" Pluto. The combined mass of Pluto and its satellite Charon is tiny: less than 5% of the mass of the next-smallest planet, Mercury. Such a feeble object may be better regarded as an extremely large comet that has lost much of its ice. Although attempts to downgrade the status of Pluto have failed — perhaps for reasons of sentiment and tradition — many planetary scientists consider that the Solar System contains only eight planets. If we do likewise, and exclude the Pluto–Charon system from the list of planets, then Earth is unique in having an exceptionally large satellite.

Note that the Moon is *not* the largest satellite in the Solar System. That honor belongs to Ganymede, which is one of the moons of Jupiter. Two other Jovian satellites — Callisto and Io — are also slightly larger than the Moon; and so is Titan (one of Saturn's moons). But Ganymede, Callisto, Io and Titan orbit giant planets. Compared to their parent bodies, these satellites are as grains of dust. Our Moon, on the other hand, is large compared to Earth: it has $\frac{1}{81}$ of the mass of our planet. The Earth–Moon system has rightly been called a "double planet." And double planets may be rare.

FIGURE 57   Pluto and Charon combined have less than 5% of the mass of Mercury, the next-smallest planet.

The Formation of the Moon

In order to estimate the scarcity of "double planets," we need to understand how the Moon formed. For many years, the formation of the Moon was one of the longstanding problems of planetary science. Three mechanisms were proposed: *co-accretion*, in which Earth and Moon formed at the same time from the gas and dust in the solar nebula; *fission*, in which Earth formed first and was spinning so quickly a large piece of material tore away and formed the Moon; and *capture*, in which the two objects formed at different places in the solar nebula, and then the Moon became trapped in orbit after straying too close to Earth. All three mechanisms had difficulties in explaining several important features of the Earth–Moon system, but it was hoped that analysis of lunar rocks brought back from the Apollo missions would vindicate one of them. Instead, it became clear that none of these ideas worked. A new theory of lunar formation was needed.

In 1975, two groups of American scientists independently proposed the *impact* hypothesis for the Moon's origin.[211] They postulated that an object with the mass of Mars struck the infant Earth in an off-center impact. The unimaginably violent collision ejected a mixture of terrestrial and impactor material into orbit around Earth, and this material quickly coalesced to form the Moon. Now, scientists generally dislike having to resort to cataclysmic or unique events to explain

FIGURE 58   Earthrise as seen on the Moon.

their observations, but at least in this case computer models can simulate possible Moon-forming collisions. Although the details of the impact are still in dispute — for example, recent work suggests that the impactor may have been more massive than previously thought — the hypothesis explains many of the observed facts about the Earth–Moon system. Furthermore, there is other evidence (from the tilts of the planets, for example) that violent collisions were not uncommon in the early Solar System. The impact hypothesis has gained a large measure of consensus among planetary scientists.

FIGURE 59   Earth and Moon: a double planet.

If our Moon was indeed the consequence of a giant impact, then the uniqueness of the Earth–Moon double planet within our Solar System need not surprise us. Although collisions Solar System objects were common, such cataclysmic Moon-forming collisions may have been scarce; perhaps the infant Mercury, Venus and Mars were simply fortunate enough to dodge the larger missiles. (It has been suggested that Venus once had a large satellite, which was formed in the same way as the Moon, but which followed a retrograde orbit: in other words, it orbited Venus in the "wrong" direction. Such an orbit could certainly occur if the satellite was created through an impact event. However, whereas tidal forces are causing our Moon to move *away* from Earth, in the case of a retrograde orbit those forces would act in the opposite direction. A satellite in a retrograde orbit moves toward the planet and is eventually destroyed. This is the fate of Triton, the largest of Neptune's satellites.) Furthermore, the Moon-forming collision occurred at a critical time. Had it happened much earlier, when Earth was less massive, then most of the debris from the collision would have ended up in space, and the Moon would have been much smaller than it is. Had it occurred much later, then Earth would have been more massive, and its greater surface gravity would have prevented the ejection of enough mass to form a large Moon.

Whereas the original scenarios for lunar formation implied that our Moon was almost a natural by-product of planetary formation, the impact hypothesis suggests that the Earth–Moon system may be exceptional. Imagine a collection of primordial stellar nebulae, each identical to the nebula from which our Solar System formed. Perhaps only 1 in 10, or 1 in 100, or 1 in 1000, would generate an Earth-like planet with a Moon as large as ours. Perhaps the figure is 1 in 1,000,000. We have no idea — and it will take huge advances in observational astronomy before we discover whether extrasolar terrestrial planets possess satellites as large as our Moon. With our present knowledge, however, it is entirely possible that Earth is unusual in possessing such a large satellite.

The Lunar Influence

Even if the Moon *is* rare, so what? If Earth were Moonless, then poets through the ages would have lost a source of inspiration. Perhaps mankind's scientific development would have been affected, since historically the Moon has played a large role in advancing our understanding of astronomy. But would life itself really have been any different?[212]

There are several ways in which the Moon exerted (or continues to exert) an influence on Earth. For example, the Moon raises ocean tides. Soon after the Moon formed it was much closer to Earth than it is now, so the tides of 4 billion years ago would have been huge — a surfer's paradise. It has been suggested that tides were a factor in getting life started, perhaps by acting as a giant mixer of the primordial soup and causing nutrient-rich pools where life may have started. This suggestion is not totally convincing, because even without the Moon we would still have ocean tides: the Sun raises tides about half as large as the present lunar tides. We would, however, miss the spring and neap tides, which depend upon the relative positions of Sun and Moon. The suggestion, therefore, cannot be ruled out.

FIGURE 60   A montage of Earth and Moon (the Moon is shown comparatively larger here than in reality).

A more subtle lunar tidal effect is its influence upon the Earth's crust. The effect of the Moon's gravity *may* have amplified volcanic activity on Earth and increased continental drift. So it is possible (though not certain) that a Moonless Earth would have been less geologically active; Earth's atmosphere, which formed by volcanic outgassing, may have taken much longer to reach the stage where life could arise. We discussed the importance of plate tectonics in the previous section.

The most important effect to consider, though, is how the Moon influences the Earth's *obliquity*. The planets all orbit the Sun in or near one plane in space; the obliquity — or axial tilt — of a planet is the angle of inclination of its equator to this orbital plane. Earth's obliquity of 23.5° gives rise to the pleasant seasons we enjoy. Other planets are not so lucky. Mercury has an obliquity of 0°, so its equatorial regions resemble Hell. Life as we know it could not survive. (Interestingly, an observer at either of Mercury's poles would see the Sun always on the horizon. Relatively little solar energy can be absorbed at the poles, and indeed the polar regions of Mercury are ice-covered.) Uranus, which has an obliquity of 98°, is almost lying on its side. One pole receives sunlight for half of the Uranian year, while the other pole is in darkness. Again, these are less than ideal conditions for life. Earth — from our biased point of view — seems to be "just right."

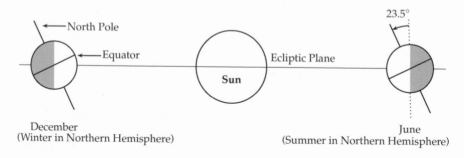

FIGURE 61   *It is Earth's obliquity — its tilt relative to the plane of its orbit around the Sun (the ecliptic plane) — that produces the seasons. For planets with a "moderate" obliquity like Earth, most of the solar energy falls in the equatorial regions, where the midday Sun is always high in the sky. The polar regions are in constant illumination for 6 months, but in constant darkness for 6 months too; even when the Sun is in the sky, it is never higher in the sky than the obliquity allows — 23.5° in the case of the Earth — so the ground is never heated really strongly by sunlight. Thus, the polar regions are cold and the equatorial regions are hot. (The figure is not to scale.)*

The impact event that formed the Moon would have caused Earth's axis of rotation to shift from its initial position. More importantly, as computer simulations have shown, the Moon plays a large role in stabilizing Earth's axial tilt over a period of many millions of years. This is important because even small changes in obliquity can cause dramatic changes in planetary climate. For example, Earth's obliquity oscillates by about ±1.5° with a period of oscillation of 41,000 years. This is only a small variation, yet it seems to be linked to the succession of ice ages that Earth has experienced over the past few million years. Mars has no stabilizing influence on its obliquity (Phobos and Deimos being merely boulders, with insufficient mass to have any influence). The obliquity of Mars is currently 25°, but this value ranges between 15° and 35°, with a period of 100,000 years. Calculations indicate that, over longer timescales, the obliquity of Mars changes chaotically: over the last 10 million years it may have ranged from 0° to 60°. With no Moon to act as a stabilizing influence, Earth's obliquity would also wander chaotically, to values as large as 90°. Even a relatively large satellite — up to half the mass of our Moon — would be unable to stabilize the obliquity; Earth requires a *large* satellite to prevent its obliquity from wandering and its climate shifting from one extreme to another.

Life on Earth has adapted well to climate change in the past, but had the Martian pattern of obliquity shifts been repeated here, it is difficult to see how advanced land animals could have prospered. Perhaps life on Earth would not have evolved into the forms we see today.

*   *   *

There are many "ifs", "buts" and "maybes" in the above discussion. We do not know whether a large satellite is *necessary* if a planet is to provide a suitable home for complex life-forms. Our own view is necessarily biased. We believe the Moon has been beneficial for the development of life here, but we do not know whether the Moon was *necessary* for life. Perhaps if we lived on a Moonless world we would be grateful we did not have one of those huge chunks of rock hanging so close to us in the sky.

And yet that nagging suspicion remains. Perhaps double planets like our Earth–Moon system *are* necessary for life, and yet they seem to form in rare, chance events. Perhaps the uniqueness of our satellite explains why we are alone. Perhaps that is the tragedy of the Moon.

SOLUTION 43: LIFE'S GENESIS IS RARE

The solution of the problem of life is seen in the vanishing of the problem.
Ludwig Wittgenstein, *Tractatus Logico-Philosophicus*

Hart's answer to the Fermi question is that life's genesis is almost miraculously rare. For practical purposes, we are alone: Earth possesses the only intelligent life — the only *life* — in the visible part of an infinite Universe.

189

This miracle loses some of its gloss in an infinite Universe: an infinite number of planets possess intelligent life-forms. However, many people find it difficult to entertain the notion of an infinite Universe with an infinite number of habitable planets. Can we not instead accept *part* of Hart's idea? Can we dispense with the astronomical notion of an infinite Universe and argue solely from biology: perhaps life is not a miracle but nevertheless arises only rarely. Maybe the Universe appears sterile because — with the exception of one or two islands of life such as Earth — it *is* sterile?

As is usual with any aspect of the Fermi paradox, there are two diametrically opposed opinions. One group argues that life is indeed difficult for Nature to create. The other argues that life is almost certain to appear on a planet as soon as conditions allow. To discuss the merits of both positions, we first need to take a lengthy detour and consider the question of what we mean by life and how life might have arisen.

*   *   *

At school, my teacher could always drive holes through the attempts of our science class to provide a definition of life. He pointed out that, by some of our definitions, fire is alive (since it grows, it reproduces itself, and so on). On the other hand, by our definitions a mule is not alive (since it cannot reproduce itself). For the purposes of this section I will try my hand at presenting another definition of terrestrial life. My old teacher could probably still drive holes through the definition, and in any case the definition might be inappropriate in the future. (In ten years, perhaps, scientists might develop a self-aware computer. Will the computer be alive? Or a century hence, perhaps, an explorer on the Altair mission will discover an evil-smelling pink crystal that every morning turns into a goo, clinging to the sides of the spaceship and eating the metal. Is the goo alive? In both cases, under my definition the answer is "no" — even though the answer should probably be "yes." We have to begin somewhere, though, and the definition given below is as good a place as any.)

I define something to be alive if it has the following four properties.

First, *a living object must be made of cells*. Every living creature on Earth consists either of a single cell or a collection of cells. If we knew how cells originated, then we might well understand how life itself originated.

There are two quite different types of cell: *prokaryotes* and *eukaryotes*. Prokaryotic cells lack a central nucleus. They are simple, small and exist in variety of types. Prokaryotic organisms are hugely successful, in large measure because their simplicity means they can reproduce themselves quickly. A recent and profound discovery is that there are two quite different types of prokaryotes:[213] eubacteria — or "true" bacteria (or, as I will write for simplicity, just bacteria) — and archaea. The two types of prokaryotic cell seem to bear no closer relationship to each other than they do to eukaryotic cells. Eukaryotic cells are much more complicated than prokaryotic cells; within an outer membrane lies a formidable array of biochemical machinery, and a nucleus enclosed within its own nuclear membranes. This complexity requires eukaryotic cells typically to possess 10,000 times more volume than prokaryotic cells. Eukaryotes are able to assemble to form complex, multicellular organisms — plants, fungi and animals.

FIGURE 62   *Four different types of archaea. (a)* Thermoproteus tenax. *Species from the* Thermoproteus *genus grow at 78–$96°C$, use hydrogen as their energy source and CO_2 as their carbon source. (b)* Pyrococcus furiosus. *"Pyrococcus" means "fireball" — a reference to both its shape and the high temperatures at which it thrives; "furiosus" means "rushing" — it can quickly double its numbers. (c)* Methanococcus igneus. *Some species grow at $85°C$ and pressures over $200\,atm$; oxygen is a poison. (d)* Methanopyrus kandleri. *Found in high-pressure ocean depths, they can survive at $110°C$.*

Thus, within the living world there are three domains: archaea, bacteria and eukarya. By this definition, viruses and prions are non-living.

Second, *a living object must have a metabolism*. Metabolism is what we call the variety of processes enabling a cell, or a collection of cells, to take in energy and materials, convert them for its own ends, and excrete waste products. In other words, all living organisms require food of some description, and all living organisms create waste. (Fire has a metabolism, as my old science teacher would point out, but we do not have to consider fire as living since it does not meet all the other criteria.) Metabolism takes place through the catalytic action of *enzymes*: without enzymes, the various biochemical reactions that take place in cells simply would not happen. In turn, enzymes are made of *proteins*. Proteins are therefore a vital constituent

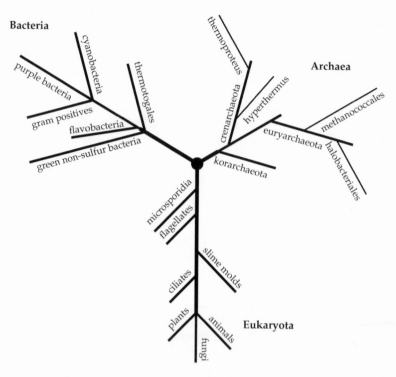

FIGURE 63 *A highly simplified sketch of the tree of life. The tree contains three domains: archaea, bacteria and eukarya. The domain of archaea contains three kingdoms: korarchaeota, crenarchaeota and euryarchaeota; the domain of eukarya contains, among others, the familiar kingdoms of animals and plants. The relationships between the three domains is controversial, and the diagram should not be taken too seriously — except that it shows that life on Earth possesses tremendous unity.*

of life — at least here on Earth. As we shall see later, the instructions for creating the various proteins necessary for a cell's existence are contained in its deoxyribonucleic acid (DNA), while the biochemical machinery of protein synthesis is based on its ribonucleic acid (RNA). In shorthand form: DNA makes RNA makes proteins.

Third, *a living object can reproduce* — or else it derives from objects that could reproduce. Cells can reproduce either individually or in sexual pairs, and the mechanism of reproduction is DNA. Clearly, then, DNA plays a central role in living organisms — just how central we will come to shortly. (Crystal structures can reproduce; however, they lack the variation that occurs when living organisms reproduce. Replication, rather than reproduction, is a better term for crystal growth, and certainly we do not need to consider crystals to be alive. On the other hand, mules and other sterile organisms came from creatures that *could* reproduce; we do not need to classify mules as non-living.)

Fourth, *life evolves*. Darwinian evolution — natural selection acting on heritable variation — is a key aspect of life.

These four properties — cells, metabolism, reproduction and evolution — are enough on which to base a discussion of life, even if the definition itself could be improved. We are now in a position to ask: how did life start?

*   *   *

It is worth stating at the outset that *nobody knows* how life started. Nevertheless, in recent years tremendous progress has been made in two directions: on the one hand, tracing life's ancestry back as far as possible, and on the other hand attempting to understand the chemical pathways that might have led to the earliest forms of life. (There is at least one other promising approach: the idea that life emerged complex and whole thanks to the self-organizational properties of chemical systems. Lack of space prevents us from discussing this approach.)[214]

The "top-down" method of looking for the origin of life is the search for LUCA — the Last Universal Common Ancestor, from which all present life must have inherited its common biochemical structures. (There is a tremendous unity of terrestrial life: all organisms, with a few minor exceptions, use the same genetic code, which enables a sequence of DNA to specify a polypeptide; all organisms use DNA to carry genetic information; and so on.) If LUCA was sufficiently simple, if it existed at a very early stage in the history of Earth — and if we can understand LUCA in detail — then we might deduce how it came to be. Unfortunately, this approach can be taken only so far. One commonly drawn picture is that LUCA was already a sophisticated organism, which had evolved considerably from the time when life first arose, before it branched into the domains of archaea and bacteria. Later, in this picture, the eukaryotic domain branched off from the archaea. This picture is complicated enough, but as the biochemical laboratories discover new information on an almost daily basis, the picture is becoming even more convoluted. We usually think of genetic information as passing only vertically — from parent to child. Early in the history of life, however, *horizontal* transfer of genes between different types of organism seems to have occurred frequently. This horizontal transfer of genetic information means that simple lineages become tangled. At the time LUCA is supposed to have existed, there may have been a pool of genes (formed from a community of cells that were able to exchange genes in horizontal fashion because they shared the same genetic code), from which the three domains arose separately. In other words, archaea, bacteria and eukarya may be equally ancient. (On the other hand, there is a suggestion that the Snowball Earth event of 2.5 billion years ago produced the conditions that gave rise to the eukaryotic cell. In other words, eukarya may be relatively *recent*; and without a Snowball Earth event they might never have arisen.) These interesting suggestions remain an active area of research.

Rather than become bogged down in the details of LUCA, we can consider the "bottom-up" approach to the question of the origin of life. We can ask: how did the universal chemicals of life — nucleic acids and proteins— come into existence? If we can understand that, then we may be able to fill in the gap between the bottom-up and top-down approaches; we may be able to understand how inanimate matter became alive.

Nucleic Acids

If any molecule deserves the title "molecule of life," it must surely be deoxyribonucleic acid — DNA. According to the definition presented earlier, life has two key aspects: it has a metabolism, and it passes on information through the reproductive process. The DNA molecule is central to both aspects. The role it plays in synthesizing proteins, which in turn allow metabolism, is described below. Here we concentrate on the reproductive aspect and briefly consider how DNA can replicate itself — while providing enough variation upon which natural selection can work.[215]

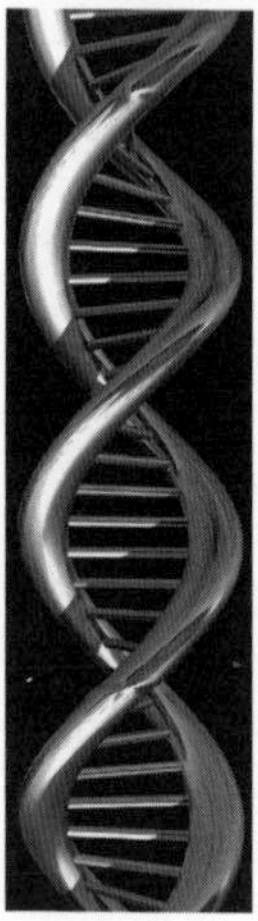

The DNA molecule is a polymer of *nucleotides*. A nucleotide consists of three parts.

First, it possesses a deoxyribose sugar. The sugar contains five carbon atoms, conventionally numbered with primes — 1' through to 5' (pronounced "one prime," "two prime," and so on). The sugar is similar to ribose, but lacks a hydroxyl molecule at the 2' position.

Second, it possesses a phosphate group. The nucleotides can link together to form long chains through so-called phosphate ester bonds — bonds between the phosphate group of one nucleotide and the sugar component of the next nucleotide. The sugar–phosphate chains form the backbone of DNA; in the familiar picture of DNA as a "ladder-like" molecule, the sugar–phosphate chains form the "rails" of the

FIGURE 64
A double helix (like DNA) as shown here in a computer- generated figure.

ladder. A chain can be indefinitely lengthened simply by attaching more nucleotides through more ester bonds; a DNA molecule can be anywhere between about 100 to a few million nucleotides in length. No matter how long the chain becomes, there are always two ends. One end has a free –OH group at the 3' carbon (the 3' end) and the other end has a phosphoric acid group at the 5' carbon (the 5' end).

Third, it possesses a pair of nitrogenous *bases*. These form the "rungs" of the DNA ladder. A base is linked to the deoxyribose sugar at the 1' carbon. A base can be either one of the purines, adenine (A) or guanine (G), or one of the pyrimidines, cytosine (C) or thymine (T). Biochemists present the nucleotide sequence in a chain by starting at the 5' end and identifying the bases in the order in which they are linked; a typical sequence of DNA may be written as –G–C–T–T–A–G–G–.

One of the key developments in science was the realization that DNA in the nuclear material of cells has two strands, twisted around each other to form a double helix, such that one strand is always associated with a complementary strand. The base G is always opposite the base C, the base T is always opposite the base A. This complementarity occurs because only these combinations of base pairs can form hydrogen bonds between them and hold the two strands together. An individual hydrogen bond is weak, but a normal DNA molecule contains so many base pairs that the two strands are held tightly together. This complementarity also means all the information is held in a single strand of DNA

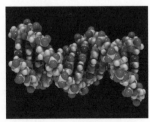

FIGURE 65   *The backbone of a DNA molecule consists of long chains of deoxyribose sugar and phosphate groups; nitrogenous bases in each helix form bonds, but they must obey the pairing rules: adenine opposite thymine, and cytosine opposite guanine.*

— and allows for the possibility of replication and reproduction.

The process of DNA replication begins when an enzyme called DNA helicase partially unzips the double helix at a region known as the *replication fork*. At the replication fork there are two strands of DNA — one of which is the *template strand*. With the bases now exposed, an enzyme called DNA polymerase moves into position and begins the synthesis of a DNA strand complementary to the template. The enzyme reads the sequence of bases on the template strand, in the direction from the 3' end to the 5' end, and adds the nucleotides to the complementary strand one at a time — always G to C and A to T. (So a sequence on the template strand of –G–C–T–T–A–G–G– would become –C–G–A–A–T–C–C– on the synthesized complementary strand, which grows in the direction from 5' to 3'.) Eventually, a complete complementary strand is formed; the DNA polymerase catalyzes the formation of the hydrogen bonds between the nucleotides on the two strands, and a new double helix can form. While this whole process takes place, a rather more complicated process manufactures a new strand that is complementary to the other original strand (or *lagging strand*). The net result is the creation of two identical copies of the original DNA double helix, and each new helix contains one strand of the original. We have a replication mechanism.

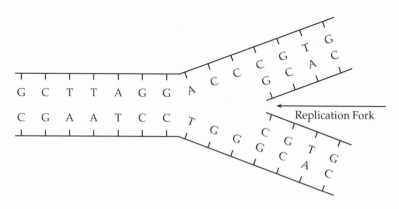

FIGURE 66   *The specific pairing of nucleotide bases — A with T, C with G — enables DNA to replicate; it is the basis of heredity. When the twin-stranded DNA molecule replicates, the two strands separate at the replication fork. Enzymes then add new bases to the two strands while following the pairing rules. The result is two molecules, both of which are identical to the original.*

(The process outlined above is a simplified version of what actually occurs. One of the aspects I omitted is the role RNA plays in the replication of DNA. Ribonucleic acid is the other major type of nucleic acid and it, too, fulfills key functions for life on Earth. There are several differences between DNA and RNA. A *structural* difference is that RNA usually appears in cells as a single chain of nucleotides, rather than as a double helix of DNA; RNA molecules are also typically smaller than DNA molecules. There are two *chemical* differences between the molecules. First, the RNA nucleotides contain the sugar ribose rather than deoxyribose (hence the difference in names between the two molecules). Second, RNA employs the base uracil (U) rather than thymine. There is also a major *functional* difference between the two acids: DNA exists solely to store genetic information in the sequence of its nucleotide bases, whereas RNA molecules *do* things. There are several types of RNA, each performing different tasks, and we shall meet three of them — messenger RNA (mRNA), ribosomal RNA (rRNA) and transfer RNA (tRNA) — below.)

The ability of DNA to replicate is the secret of life's ability to reproduce. This ability explains why offspring look like their parents — snakes beget snakes, woodpeckers beget woodpeckers, and humans beget humans. But for life to *evolve*, and for species to change into other species, heredity must be imperfect. There must be some variation among offspring: natural selection cannot adapt things that do not vary. Fortunately, there is variation when DNA replicates. From time to time, a *mutation* occurs: there is a change in the sequence of nucleotide bases. These mutations occur randomly from radiation damage, from chemical agents and simply from errors in the DNA replication process. (The rate of mutation is remarkably small, due to various checks that take place when DNA replicates. After the first stage of replication there are two error-correcting stages: *proofreading* and *mismatch repair*. These extra stages minimize the error rate to 1 in 10^9.) If an error occurs in a part of DNA that codes for a protein (more on this below), then the mutated DNA will produce a different protein. If the protein performs its intended job better than the original, then the mutation will be beneficial for the organism (and perhaps increase the probability of the organism's survival and thus, through increased numbers of offspring, of its own continued existence); but more probably, the mutation will be harmful or at least neutral. The point is that mutations give natural selection something on which to work.

If all that nucleic acids did was replicate, then they would be only marginally more interesting than self-replicating crystals. While DNA can *store* genetic information, it would be of little use if the information was not retrieved and put to use. It would be like having a public library stacked full of books, but with no one allowed to read any of the volumes. What makes nucleic acids so fascinating is that they code for and construct proteins. And proteins are what make life so interesting. Proteins enable life to *do* things.

Proteins

Proteins are complicated macromolecules that exhibit tremendous versatility. They function as enzymes (which make possible a cell's metabolism), they act as hormones (thus providing a regulatory function; insulin is a common example), and they provide structure (our fingernails, hair, muscles, and the lenses in our eyes are all proteins).

A protein is a long sequence of *amino acids* folded into a three-dimensional structure. A particular sequence of amino acids folds into a particular structure. Change the sequence and you change the way the protein folds up — and thus the task that the protein can fulfill, since the biochemical task that a protein can carry out depends critically upon its shape in three dimensions. Proteins make use of twenty different amino acids. There are many other amino acids in Nature, and several of them are important in biology; but proteins use only twenty. All the amino acids have a common structure: an amino group (H_2N), a residue or R group (CHR) and a carboxyl group (COOH). The general structure is written H_2N—CHR—COOH, and the chain forms by linking the amino end to the carboxyl end by peptide bonds. (A chain of amino acids is thus called a *polypeptide*; a protein is simply one or more polypeptides.) What makes each amino acid unique is the R side chain: different amino acids have different R groups and thus possess different properties. For example, some side chains create an amino acid that is hydrophobic; such amino acids tend to cluster on the inside of a protein and thus play a factor in determining the three-dimensional structure of the molecule. Other side chains make an amino acid that is hydrophilic — in other words, it reacts readily with water.

FIGURE 67 *The* ras *protein, which acts as a molecular switch governing cell growth. Knowing the structure of this protein in three dimensions may enable scientists to devise methods of turning off the switch in cancer cells. However, computing the way in which a sequence of amino acids will fold is an extremely difficult problem.*

Each amino acid is coded for by a set of three RNA nucleotide bases called a *codon*. Since there are four bases (A, C, G, U) there are $4 \times 4 \times 4 = 64$ codons. In theory, then, codons could code for 64 amino acids — and yet only 20 different amino acids are used in protein synthesis. The *genetic code* is thus degenerate: 3 of the codons represent an "end of chain" command, and the other 61 codons code for the 20 amino acids. In other words, nearly all amino acids are coded for by several codons. (For example, the amino acid cysteine is coded for by the codons UGU and UGC; isoleucine is coded for by the codons AUU, AUC and AUA; and so on.) The genetic code is essentially universal: with only minor exceptions, all organisms on Earth use it. (Does the universality of the genetic code imply that it is the only possible code? Perhaps there were originally several different codes, and this one just happened to win out over the others. But if the present uniqueness of the code means that it arose only once in the history of life, perhaps the development of an effective code represents a difficult barrier for evolution to overcome.)

The way a cell goes about synthesizing a protein is at once wonderfully simple and marvelously intricate. A highly simplified version of the process proceeds as follows.

The information on how to build proteins — and thus an organism — is contained in the organism's DNA. First, then, when a cell receives a signal asking for it to produce a certain protein (and let us suppose the protein is a single polypeptide), the double-helix of DNA unzips in the region of the *coding strand*. This is like the template strand mentioned above and contains information for that particular protein. A region of DNA that codes for a polypeptide (or, more accurately, that codes for some form of RNA) is known as a *gene*.

An mRNA copy of the gene is made in a *transcription* process — so called because each triplet in the DNA strand is transcribed into the corresponding codon in mRNA. The mRNA then moves from the nuclear material to the cytoplasm of the cell, taking with it its information on amino acid sequences. Within the cytoplasm, organelles called *ribosomes* take the mRNA and use the information contained in the codon sequence to synthesize the protein, adding amino acids onto the growing chain. This process is called *translation*, since a ribosome uses the genetic code to translate from the sequence of codons into a sequence of amino acids. A key ingredient here is tRNA — small molecules, each of which can bind only to a particular amino acid. A series of enzymes is required to catalyze the binding process; each enzyme recognizes one particular tRNA molecule and the corresponding amino acid.

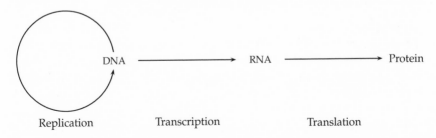

DNA   ⟶   RNA   ⟶   Protein

Replication   Transcription   Translation

FIGURE 68   *The DNA molecule stores genetic information, and replicates that information when a cell divides. The expression of that genetic information does not take place directly. Instead, DNA is first transcribed into RNA. Information stored in the "four-letter" alphabet of nucleotides (the alphabet used by RNA) is then translated into the "twenty-letter" alphabet of amino acids (which are used to construct proteins). The Central Dogma of biology, first stated by Francis Crick, is that the information flow follows the direction of the arrows in this diagram. In particular, RNA can synthesize proteins through translation, but reverse translation never occurs.*

Protein synthesis always begins with methionine (with codon AUG) and continues until the ribosome encounters one of the stop codons (UAA, UAG or UGA), at which point the protein is released and the synthesis is over. (This provides an outline sketch of protein synthesis, at least for prokaryotic cells. In eukaryotic cells, the process is further complicated by the presence of sequences of DNA that do not code for anything. A further step is required to remove this seemingly useless information. Space here is too limited to go further into the details of protein synthesis, but there are many excellent sources available for further reading,[216] and fortunately we do not need extra detail to continue the discussion.)

To recap: DNA stores genetic information and replicates the information when a cell divides. That is all it does. The messy business of actually *expressing* the information is left to the more versatile RNA; using the universal genetic code, information is transcribed from DNA into RNA and then translated into protein synthesis.

How Did the Ingredients of Life Arise?

Let us assume, for the moment, that the many intricate steps leading from the first proteins and early nucleic acids through to LUCA are, if not inevitable, at least capable of being understood using well-known physical and chemical processes. We are still left with the question: how did the first proteins and nucleic acids come into existence? If the step from inorganic chemistry to DNA and proteins is a rare phenomenon, then we have a resolution of the Fermi paradox. For without these large molecules, evolution cannot begin the step to LUCA and then to the variety of life we see around us. Life, at least as we know it, cannot exist.

The basic building blocks of the vital macromolecules appear to be easily synthesized. We find amino acids, for example, both in interstellar space[217] and in experiments that attempt to mimic the chemistry of early Earth.[218] In 1953, Stanley Miller performed a classic experiment in which he passed an electric discharge through a vessel containing a mixture of water, methane and ammonia. The experiment was intended to investigate the effects of electric currents passing through the atmosphere of the early Earth. At the end of his experiment, Miller found many organic compounds in the vessel. Other scientists have disagreed with Miller's choice of model atmosphere, but the results were unarguably dramatic. It seems probable that amino acids could have formed on Earth soon after our planet cooled; amino acids are almost an inevitability of organic chemistry and the marvelous associative properties of carbon. Similarly, sugars, purines and pyrimidines — the components from which nucleic acids develop — can form in Miller-type experiments (although it must be admitted that yields are often low).

Although the details have yet to be determined, we need not suppose that the basic chemical building blocks required for life are in any way exceptionally rare. We can be less confident, however, about the probability of natural processes successfully linking these components into the molecules of life — nucleic acids and proteins. Indeed, it is at this point many creationists (and a few scientists) claim life on Earth is unique: they argue the probability of random processes creating a nucleic acid or a protein is tiny.

Consider, for example, serum albumin (an average-sized protein produced in the liver and secreted into the bloodstream, where it performs several necessary tasks). Serum albumin contains a chain of 584 amino acids, which are curled up into a sphere. In our bodies, the synthesis of the molecule is under the direction of nucleic acids. But imagine a time before DNA existed, so that a molecule of serum albumin had to be synthesized by adding one amino acid at random to the end of a growing chain. The chances are negligible — just 1 in 20^{584} — that random processes would produce the protein. Similarly, "genesis DNA" — a primitive chain of nucleotides that some scientists propose as being necessary for life to start — has a low probability of being created by chance.[219]

Making a Protein Through Random Processes

Since there are 20 amino acids from which to choose, at each step the probability that the correct amino acid is chosen to add to the end of a growing chain is 1 in 20. Therefore, for serum albumin, which has 584 amino acids, the probability that every amino acid is chosen in the correct order is 1 in 20^{584} — which is the same as 1 in 10^{760}. This is an incredibly small probability. There is essentially zero chance that this protein can be synthesized by such a random process. Even a small protein like cytochrome c, which consists of just over 100 amino acids, has only a 1-in-10^{130} chance of being synthesized at random. Again, for practical purposes, this number is indistinguishable from zero.

The beginning of life seems to suffer from a "chicken and egg" paradox: DNA contains the instructions necessary for the assembly of amino acids into proteins, but every DNA molecule requires the help of enzymes (in other words, proteins) to exist. DNA makes proteins makes DNA makes proteins. Which came first?

Although these criticisms seem to be fatal to the claim that life arose by chance, biochemists have in recent years made great progress in countering them. The details are not yet complete, but there is no reason to suppose the problems are insurmountable. Begin with the combinatoric arguments against the primordial synthesis of proteins. There is indeed essentially no chance of cytochrome c, for example, somehow coming together by accident. But if we allow for a period of prebiotic *molecular* evolution, then proteins could be synthesized through the workings of chance.

For example, imagine a lake somewhere on the still-young Earth. Suppose that in this lake there were only 10 different amino acids capable of forming peptides; and suppose that a peptide with a length of 20 amino acids displayed some catalytic function making it favored by natural selection. Then Nature only needed to try out 10^{20} combinations to hit on this peptide — still an enormous number, but a number that *could* comfortably be accommodated in the timescales available. Once the peptide was created, natural selection would ensure the amount of peptide in the lake increased in volume. Suppose that 1000 different "useful" peptides, each 20 amino acids in length, were created in the lake. If two such peptides could join to form a single chain, then 1 million different peptides with a length of 40 amino acids could be formed. Again, Nature would have plenty of time to try out all the combinations. In the same way, peptides containing 60 amino acids could be synthesized, and 80, and 100 ... in short, there *was* time for proteins to arise in that ancient lake. And there were many millions of lakes on the early Earth. (The particular proteins that arose would surely have been an historical accident. Replay the tape of history, and the proteins we use might be very different.)

Similar sorts of argument involving prebiotic molecular evolution can be used to counter the claim that "genesis DNA" was a miraculous fluke. However, such arguments may be unnecessary. It seems increasingly plausible that the original self-replicating molecule was not DNA, but one of the varieties of the much simpler RNA molecule. Furthermore, RNA provides an answer to the "chicken and egg" paradox. In the early 1980s, Sidney Altman and Thomas Cech demonstrated that some types of RNA molecule could also act as catalysts; they could play the role of enzymes. These RNA enzymes — or ribozymes — led to the idea of the "RNA world" — a time in the early history of life when catalytic RNA enabled all the chemical reactions to take place that are necessary for primitive cellular structures. In a sense, neither the chicken nor the egg came first: catalytic RNA acted both as genetic material and as enzymes.[220]

There seems to be no fundamental reason to suppose that the basic molecules of life could not arise through natural processes that had a reasonable chance of occurring. (Although, in all honesty, one has to concede that the chemical pathways leading to the first RNA molecules are still murky. The subsequent evolution of cellular structures up to LUCA is just as unclear. There are several competing scenarios, each with their advantages and drawbacks. Furthermore, several questions — such as why life uses only the left-hand form of amino acids, and whether the genetic code is inevitable or simply one of a whole raft of possible codes — are outstanding.

But progress in these fields is rapid, and we can expect the picture to have more clarity within a few years. Even if life turns out to have a completely different origin from that sketched above — and there are several other competing hypotheses — we are not yet driven to the hypothesis that life was some bizarre fluke.) There is, however, one last argument to consider regarding the probability of the early Earth being the site of the genesis of life: paradoxically, life seems to have arisen here *too* easily!

FIGURE 69   *Stromatolites, similar to the one pictured here, are the oldest known fossils. The oldest of them are 3.5 billion years old.*

When Did Life Arise on Earth?

Life appears to have had little trouble in emerging on Earth. We know our planet formed about 4.55 billion years ago. A maximum of 700 million years after the formation of Earth — 3.85 billion years ago — it seems that life had evolved. We believe this to be the case because certain sedimentary rocks in Isua, Greenland — rocks that are among the oldest on this planet — contain isotopes of carbon in a ratio that is a sign of biological processes.

(The interpretation of these measurements is not without controversy. It may turn out that non-biological processes can generate a similar isotopic ratio of carbon. Nevertheless, many biologists accept that life was in existence at this time.) Since these are among the *earliest* known rocks, we can say there is little direct geological evidence for there ever being a time when life was absent from Earth! The earliest fossils are not much younger than the Isua rocks; stromatolites — mounds built up of layers of cyanobacteria and trapped sediment — are preserved as fossils in the Warrawoona Group in Western Australia. These stromatolites are 3.5 billion years old.

The haste with which life arose is almost too quick for comfort. The timespan mentioned above for the emergence of life, namely 700 million years, is an upper limit: that time span is squeezed from both ends. On the one hand, there was presumably some evolutionary process leading to the life-forms we find in the Greenland rocks; certainly the cyanobacteria of the Warrawoona Group had a biochemistry as sophisticated as other forms of life. In other words, if we found older rocks we might well find evidence for life in those rocks — perhaps simpler forms of life, but life nonetheless. Thus, life almost certainly emerged *before* Earth was 700 million years old. On the other hand, life could not have survived conditions on the very early Earth. (The initial period after formation of Earth, some 4.55 to 3.9 billion years ago, is called the Hadean era. The Isua rocks were laid down in the Early Archean era, which runs from 3.9 to 2.9 billion years ago.) As discussed on page 186, the early part of the Hadean era saw Earth peppered with impacts from large bodies. It is difficult to comprehend the violence of the literally Earth-shattering impact that gouged out the material that became our Moon. Certainly the impact would have sterilized the Hadean Earth: if any form of life was in existence before the impact, it could not have survived. So the period of 700 million years postulated for the emergence of life is an upper limit: the actual period was probably less than this.

Although several hundred million years may seem to offer plenty of time for life to evolve, it is worth remembering that the gap between life and non-life is huge, and that evolution can be a slow process. As the biologist Lynn Margulis famously put it: "The gap between non-life and a bacterium is much greater than the gap between a bacterium and man." And yet this gap was bridged relatively quickly. Some scientists find it difficult to accept that life could have begun so early on Earth without help, and have resorted to the panspermia hypothesis (see page 44).

If life indeed came to Earth through space, then there are implications to be considered for the Fermi paradox. The implications, however, depend upon exactly where the seeds of life came from. If life traveled through interstellar space and seeded our planet, then presumably there are countless numbers of planets in the Galaxy that were similarly seeded. Life will be everywhere. On the other hand, some astrobiologists have suggested that life originated on Mars — where conditions may have been more conducive to the development of life — and was transported to Earth on rocks that were ejected into space following impact events. As Mars lost its water, life died there; as conditions on Earth became more settled, life flourished here. If this is what happened, then life may be scarce even if life itself forms easily. It could be that *two* planets are required for life to prosper: a small planet on which life can originate, a nearby, more massive planet that can provide a long-term home for life, and meteorite impacts generating sufficient ejecta to transport life from one planet to the other. Such a combination of circumstances could be highly improbable.

Finding Life on Other Worlds

There is, of course, a direct way of determining whether life can arise under natural conditions: we could try to find it on other planets.[221] The SETI activity is one way of doing this, but there is another way. We could look for primitive life elsewhere in the Solar System. If we found life elsewhere — even the simplest microbe — we would at least know that life is not unique to Earth. Finding life on other worlds would almost certainly tell us something about how it arose on this one.[222]

The key ingredient of life seems to be water: find water and there is a chance of finding life. We know that, in the past, Mars almost certainly possessed water; so there is a chance — no matter how remote — of finding fossil remnants of past Martian life. In the present-day Solar System, at least three bodies besides Earth might have oceans. Two of the moons of Jupiter — Europa and Callisto — might possess subsurface oceans of water. These bodies are far from the warmth of the Sun, of course, and on the surface of these moons there are thick sheets of ice; but geothermal and tidal heating may be enough to maintain liquid water deep beneath the surface. Titan, a moon of Saturn, may possess a subsurface ocean of ammonia-water. Here are three places that maybe — just maybe — are home to alien life. It would not be life with which we could communicate; but if we knew that life arose *independently* in our Solar System more than once, then how could we reasonably argue that life is rare throughout the Galaxy? Surely, then, missions to Europa and Callisto, and later to Titan, should be a priority.[223]

FIGURE 70   *If there is an ocean beneath the ice of Europa, then a hydrobot similar to this artist's impression will probably be used to explore it. NASA scientists are currently examining the details of how to send a hydrobot to Europa, have it penetrate the ice and reach the ocean without introducing contamination, and then have it send back information to Earth.*

SOLUTION 44:   THE PROKARYOTE–EUKARYOTE TRANSITION IS RARE

Life may change.
Percy Bysshe Shelley, *Hellas*

For some biologists, the haste with which cells appeared on Earth implies that the generation of life from inanimate matter is straightforward. If Earth is typical, then millions of planets in the Galaxy may be home to microbial life. However, although the eukaryotes might be as old as the archaea and bacteria, the byzantine biochemical machinery of the modern eukaryotic cell took a long time to reach its present level of sophistication. It may have taken a billion years; maybe longer. The development of large multicellular organisms took longer still. This is not necessarily surprising: eukaryotic cells are immensely more complex than prokaryotic cells, and

several evolutionary developments had to be made before different eukaryotic cells could learn to cooperate and function effectively in groups.
But perhaps this long time implies that the development of the eukaryotic
grade of life follows a tortuous, difficult path. Presumably, complex multicellular life anywhere in the Galaxy must evolve from single-celled microbial life. Maybe complex eukaryotic life — and thus life capable of communicating over interstellar distances — has not yet developed on other planets. Perhaps this explains the silence of the Universe. Perhaps the Galaxy
is filled with planets on which life has stalled at the prokaryotic stage.

What led to the change from the prokaryotic grade of life, which dominated life on Earth for so long, to the eukaryotic grade of life we see all
around us today? To answer that — and to attempt to understand whether
the eukaryotic grade of life might be a rare phenomenon — we need to
understand something of the differences between two types of cell.

Differences Between Prokaryotic and Eukaryotic Cells

Whichever way you consider it, bacteria have always been the most successful life-forms on Earth. Their simplicity, combined with their capacity
to reproduce quickly, almost guarantees success. They evolve biochemical responses to environmental challenges, so even though they all tend to
look alike, different bacterial species possess different metabolisms and can
inhabit a wide variety of niches. They are also extremely hardy, and some
species seem to have survived unchanged for billions of years.

Complex eukaryotic life-forms such as plants and animals are much
less robust. They are prone to mass extinctions, and even in the natural run
of things the typical lifespan of an animal species is measured in millions
rather than billions of years. Nevertheless, the eukaryotic grade of life is
much more interesting than the prokaryotic grade. Eukaryotes evolve mor-
phological responses to environmental challenges — in other words, they
develop new body shapes and body parts — which leads to a variety and
freshness absent in the prokaryotes.

A major difference between eukaryotic and prokaryotic cells is that the
latter have rigid cell walls or very rigid cell membranes, whereas eukaryotic cells either lack cell walls or have very flexible walls. This flexibility
allows eukaryotic cells to change shape, and also to engage in *cytosis* — a
process wherein the cell membrane pushes inward to form an intracellular
vacuole. Many cellular processes employ cytosis, but perhaps its main role
is in *phagocytosis*. In phagocytosis, a eukaryotic cell engulfs a particle of
food into a food vacuole, where enzymes then digest it. Obtaining nour-
ishment like this by predation is a much more efficient process than that

employed by bacteria, which secrete digestive enzymes into the surrounding medium and then absorb the resulting molecules.

Another distinguishing characteristic is that a eukaryotic cell has a *nucleus*, separated from the cytoplasm by two membranes, which contains the cell's DNA. Eukaryotic cells also contain *organelles* — little organs — which are separated from the cytoplasm by membranes. The organelles include the *mitochondria* (which play a vital role in energy metabolism) and the *plastids* (which play a role in photosynthesis in plants and algae). In the early 1970s, Lynn Margulis argued that organelles must have arisen by symbiosis. She reasoned that, billions of years ago, very primitive eukaryotic cells would have used phagocytosis to ingest smaller prokaryotic cells for food. Some prokaryotic cells might have been indigestible and would have stayed in the larger eukaryotic cells for some time. And some of those prokaryotes would have performed functions — such as the transformation of energy — more efficiently than their hosts. Both cells would benefit from partnership — and both would have a selective advantage when it came to passing on their genes. An initially indigestible bit of food would become indispensable to the smooth running of a eukaryotic cell. Support for Margulis' idea has come from DNA sequencing. Mitochondria and plastids have their own DNA, which is different from the DNA in a cell's nucleus. It turns out that mitochondrial DNA and plastid DNA are much closer to prokaryotic than eukaryotic DNA. The mitochondria, for example, probably share a closest common ancestor with present-day symbiotic purple non-sulfur bacteria. (Direct evidence for Margulis' hypothesis has probably been erased by a billion years of evolution, but the hypothesis makes so much sense that it is widely accepted.)

Another major difference exists between the two cell types. Unlike prokaryotes, new eukaryotes can form through the fusion of gametes from two parents; in other words, sex can occur. Furthermore, the amount of genetic information stored by eukaryotes (and passed on either through sex or through parthogenesis) is far greater than that stored by prokaryotes.

Finally, eukaryotes possess a *cytoskeleton*. The cytoskeleton consists of actin filaments, which resist any pulling forces that might act on a cell, and microtubules, which resist any shearing or compression forces that might act on a cell. Thus, even in the absence of a rigid cell wall, a eukaryotic cell can maintain its shape and integrity. But the cytoskeleton can do much more: it can draw the cell into a variety of temporary shapes, it marshals the organelles into various positions, and it allows the eukaryotic cell to increase in size. Actin and tubulin — the structural proteins from which the cytoskeleton forms — are thus among the most important of all proteins for the development of complex life.

How Probable Was the Development of Eukaryotic Cells?

Was it inevitable, this transition from a primitive cell to the awesome complexity of a modern eukaryotic cell? Or was it a fluke? These are difficult questions to answer, not least because the many steps involved in the transition occurred so long ago. One of the first steps must have been loss of the rigid cell wall, even though this would have been fatal to most organisms that attempted it. (Penicillin, for example, works by blocking the formation of bacterial cell walls. Without a rigid wall to protect them, most single-celled organisms are vulnerable to attacks from the environment.) Disposing of the cell wall was *ultimately* extremely useful, because it enabled phagocytosis to occur. But phagocytosis evolved at a later date and thus could have provided no *immediate* benefit to the organism that lost the wall. Evolution has no foresight; unless an organism can survive in the here-and-now and pass its genes on to offspring, any potential it may possess will be lost. Somehow, in ways not yet understood, some organism managed to employ new structural proteins — actin and tubulin — and develop a cytoskeleton that helped mitigate the loss of the wall. How likely was this to happen? We simply do not know. The origin of organelles is better understood — it came about by symbiosis as, perhaps, did the cell nucleus — but what about the origin of what may be the most important innovation of all: cooperation between cells?

Multicellular Organisms

A few prokaryotes have adopted a multicellular way of life. Stromatolites, for example, consist of bacterial colonies. In general, though, prokaryotic cells live a solitary life (and even in the case of stromatolites it is debatable whether the term "organism" is warranted.) For most of Earth's history, eukaryotic cells also lived isolated lives. Then a remarkable transformation occurred. Some eukaryotic cells discovered the benefits of joining together. Because the cells had no external walls isolating them from the environment and from each other, they were free to exchange information and to share materials. The result was the world we see today: three kingdoms of organisms that are hugely complex and various — fungi, plants and, most complex of all, animals.

What caused eukaryotic cells to pool their resources is not known. It is not even entirely clear *when* the switch to multicellularity occurred. A crucial event in the history of life was the Cambrian explosion 540 million years ago, which saw the various animal body plans laid down, and which seems to have been a key step on the path to intelligent life on Earth. The

Cambrian explosion saw the fossilization of a broad assortment of animals — so animals were certainly in existence at that time. There are few, if any, animal fossils in rocks older than 540 million years. However, all we can deduce from this observation is that large animals with hard body-parts became common in the Cambrian period. It is entirely possible that small soft-bodied animals were in existence before the Cambrian period and died leaving no trace. (Nematodes are perhaps the most abundant type of animal in the world today. They must have existed since at least the Cambrian explosion, yet they have left no trace in the fossil record.) Gene sequencing leads some biologists to believe animals originated 1 billion years ago, which, if true, means the fossil record relates to only half of the history of animal life on Earth. Whether animals originated a billion years ago, half a billion years ago or some time in between, the fact remains they are johnny-come-latelies in the history of Earth. Single-celled creatures had been around since soon after the Earth cooled; it took 3 billion years for complex creatures to develop. Why the long wait for multicellularity?

One (still controversial) suggestion is that a rise in the oxygen content of the atmosphere ignited the Cambrian explosion.[224] Early in Earth's history there was essentially no free oxygen. This lack of oxygen posed no hardship for primitive prokaryotes; indeed, for the first living organisms, and even for some present-day bacteria, exposure to oxygen meant certain death. However, organisms such as cyanobacteria produced oxygen as a by-product of their metabolism. For 2 billion years — from about 3.7 billion years ago to about 1.7 billion years ago — these organisms pumped oxygen into the environment. For most of that time there were enough sinks, such as iron dissolved in the oceans, to trap the oxygen. Eventually, though, the sinks became full — and the oxygen content of the atmosphere began to rise. For many organisms, this event spelled doom; the "oxygen crisis" must have created the biggest of all mass extinctions, with many prokaryotic species simply failing to adapt to the large-scale release of such poison. Some organisms, though, prospered: they evolved a metabolism based on oxygen, breaking down food into carbon dioxide and water. This oxygen metabolism generated more energy than did the anaerobic metabolisms, and the organisms prospered; the eukaryotes prospered most of all. Even until about 550 million years ago, however, the concentration of oxygen in the atmosphere and dissolved in the oceans was far less than present-day amounts. Any animals existing before this period must have obtained oxygen for their tissues by diffusion, which is a slow process. Those animals would have had no heart — at least, no pump — nor would they have possessed a circulatory system. They would have been small, gossamer-like creatures, so it is small wonder that they left no trace in the fossil record. But then, for some reason that is not entirely clear, the atmospheric oxygen level rose yet again in the Cambrian period. Several key evolutionary developments took place — gills, haemoglobin in blood, hearts — allowing marine animals to make much more efficient use of oxygen and to transport the gas to different tissues. Animals became bigger and bulkier and were able to develop various specialized organs. Perhaps the emergence of a predator caused other species to evolve protection in the form of hard shells — and finally animals could become fossils.

The suggestion, then, is that the Cambrian explosion was caused by a rise in the level of oxygen in the atmosphere. And maybe this was a less-than-inevitable occurrence. Perhaps on most planets the development of large multicellular organisms does not take place.

*   *   *

As we have seen, there were many steps leading from simple unicellular organisms to complex organisms consisting of groups of cells working together. On Earth, it took billions of years for these steps to occur and for animals to appear. Which of these steps were vital and the timescale for these steps to occur are still a matter for debate. And it may be that some of the steps required environmental rather than biological changes.

It is at least a plausible resolution of the Fermi paradox that life elsewhere in the Galaxy has stalled at the unicellular stage. We may one day visit planets and find everywhere oceans teeming with strange, microscopic organisms — lots of life, but life at a low grade. Perhaps nowhere else did the right sequence of biological and environmental events take place that would make possible the evolution of animal life — and thus intelligent species with which we can communicate.

SOLUTION 45: TOOLMAKING SPECIES ARE RARE

Man is a toolmaking animal.
Benjamin Franklin (attributed by James Boswell, *Life of Johnson*)

The road from the first eukaryotic cells to the animals we see today was tortuous and, many would argue, far from inevitable. There might be several hurdles to overcome before animals species can flourish, and perhaps the answer to Fermi's question lies in those hurdles. But let us suppose that once the eukaryotic cell has developed, then it is all downhill from there; given enough time, advanced animal life will definitely appear on a planet. Does it then follow that an animal species capable of building a radio telescope will develop? Maybe not.

Chapter 5

People have long sought to identify one defining characteristic of mankind — one attribute distinguishing *Homo sapiens* from the animals on Earth. A trait often proposed for this role is tool use and toolmaking. "Man the Toolmaker" is a powerful image. If toolmaking is unique to humans, if among the billions of species that have ever lived on Earth *Homo sapiens* alone has mastered the intricacies of tools, then we might have a resolution of the Fermi paradox. Perhaps tool use and toolmaking are rare *anywhere* in the Galaxy. And without tools to build spacecraft or construct beacons, it is presumably impossible for a biological species to make their presence known across the depths of space.

*   *   *

There is a major difficulty with this suggestion: many species use tools and quite a few species make them.

For example, several species of birds use twigs to pry out grubs from the bark of trees. Sea otters place anvil stones on their chests and use them to smash open crab shells. Wasps use small pebbles to help hide the entrances to burrows where they have laid eggs. Egyptian vultures pick up rocks in their talons and drop them on ostrich nests to crack open the eggs. The list of tool use among animals is a long one. Of course, none of these examples are what *we* understand by tool use. These animal behaviors are all highly stereotyped; they are specific, repetitive responses to particular problems. Change the nature of the problem and these creatures are lost. Nowhere do these animals display insight; those elaborate displays are the intelligent result of brainless evolution.

If we require better examples of tool use, then we are forced to look at the primates. At this point *Homo sapiens* begins to seem special, if not unique, for even among the primates there are relatively few "real" examples of tool use. Apart from the great apes, which we will come to in a moment, the only primate that spontaneously uses tools in the wild is the capuchin monkey (the type of monkey employed by organ-grinders). Field workers have observed capuchins put stones and sticks to a variety of uses; among other things, the monkeys use them to obtain food and repel predators. In laboratory settings, capuchins learn to use sticks to obtain nuts from different experimental setups. However, capuchins have no real understanding of the *principles* of tool use, nor any comprehension of why a particular technique might work or fail. Watch them, and it is clear that they engage in trial-and-error prodding and poking.

Of all the animals, it is the chimpanzee that seems to make the most creative use of tools in the wild. The chimpanzees of West Africa, for example, use a hammer stone and an anvil stone to crack open nuts (and they make a better job of cracking nuts than I do at Christmas). Suitable stones can be in short supply, and the chimpanzees often have to carry them over long distances to a source of nuts. These chimps plan ahead. The chimpanzees of Tanzania use a variety of twigs for a variety of purposes, and the twigs are modified beforehand if necessary. These chimps are *making* tools. They also employ various items of foliage for a variety of functions — banana leaves are used as umbrellas, smaller leaves are used to wipe off dirt, and chewed leaves are used as sponges. Perhaps even more impressive is the achievements of Kanzi — touted by some as a veritable Edison of the animal kingdom. Kanzi is a bonobo (a species that, along with its sibling species, the chimpanzee, is our nearest relative in the animal kingdom). Among many other accomplishments, Kanzi has mastered the rudiments of stone tool production. (This particular accomplishment should not be oversold, however. Kanzi was taught how to take rock cores and from them make stone flakes capable of cutting a cord. After about one year, Kanzi had spontaneously made several improvements and advances to the flake-making technique it had been taught. The stone flakes it produced were small items, however; Kanzi clearly had no understanding of the properties of rock and no insight in how best to fracture rock to obtain large, useful flakes. Furthermore, bonobos have never been observed to use tools in the wild. Kanzi had the benefit of intensive training and teaching by humans.)

FIGURE 71 *These mesolithic flint tools — small blades and a scraper — are 9000 to 8500 years old. Their construction is quite beyond the abilities of animals.*

The lesson to be learned from these examples is perhaps this: animals use tools because they can. Tool use is less an indicator of the natural "intelligence" of an animal than a reflection of manipulative abilities (and the evolutionary adaptations its species has made to fit a particular ecological niche). A bird can use its beak for a variety of purposes, an elephant can use its trunk, and a chimpanzee is fortunate in possessing a hand that can manipulate objects in several ways. However, a camel, or a cow, or a cat, is never going to be a natural tool user — not because these creatures are inherently inferior to birds or less intelligent than chimpanzees, but simply because they lack the requisite manipulative ability. Presumably if they *could* use tools, they would.

Mankind is fortunate: our species possesses a hand that permits a quite astonishing range of actions. (Count how many different ways you configure your hand to carry out tasks during a typical day. You will be surprised.) We are excellent toolmakers because we have the manipulative abilities to be excellent toolmakers — and when this is combined with our other traits, such as language and social living, it is not difficult to understand why our use of tools is qualitatively different from that of other species. (The view I have described above is rather different from the traditional view, which says we are better toolmakers than other animals because we are more intelligent than other animals. But one can make a strong case for saying that early man's use of tools was one of the drivers of increasing human intelligence — intelligence that was then co-opted for other purposes. The neuronal circuitry required to control the precision manipulations of the human hand, and to govern activities like the throwing of projectiles at moving prey, is phenomenal — and quite beyond the capacity of any present-day robot.)[225]

We have to ask, then: what is the chance that an extraterrestrial species will follow the same sort of evolutionary route that man followed? Of course, an extraterrestrial does not need five-fingered hands in order to build a radio telescope; the course of evolution does not have to be identical. But in order to develop advanced technology it will need *some* sort of precision-manipulative ability (whether using claws, tentacles or something beyond our imagination) combined perhaps with other characteristics such as stereoscopic vision. We have no way of knowing how probable or improbable such an evolutionary outcome would be. But I for one find it difficult to believe that *no* other species could have evolved the requisite toolmaking abilities. Toolmaking is perhaps one more hurdle that has to be overcome before a species can communicate, yet one more way in which a world full of life can still fail to produce a civilization capable of communicating with us. But surely this cannot be the sole explanation of the Fermi paradox.

SOLUTION 46: TECHNOLOGICAL PROGRESS IS NOT INEVITABLE

Progress, man's distinctive mark alone.
Robert Browning, *A Death in the Desert*

Man is now the only hominid species on Earth, but until recently — until about 30,000 years ago — we shared the planet with at least one other human species. We certainly co-existed with *Homo neanderthalensis*, and we may have co-existed with *Homo erectus*. (30,000 years seems a long time, but it is a mere instant in the Universal Year; even in the history of our species it represents less than a third of the time we have been in existence.) This realization — that once we were not alone — is quite recent, as many anthropologists used to think only one species of hominid could have existed at any one time; in this view the Neanderthals must have been our ancestors. Recent evidence, however, seems to rule out this possibility. Studies of mitochondrial DNA from *Homo sapiens* and *Homo neanderthalensis* show that they were two genetically distinct species. The finding is backed up by recent computer reconstructions of the skulls of Neanderthals and early modern humans: skull development was quite different. So it seems certain that *Homo sapiens* and *Homo neanderthalensis* are separate species, sharing a common ancestor in the distant past — perhaps as long ago as 500,000 years — before evolving in separate ways. It seems equally clear that, although there may have been a small degree of interbreeding, the Neanderthals contributed nothing to the modern human gene pool.[226]

Earth may have been home to 20 or more hominid species at various times, and some of these species must have co-existed. The simple picture of hominid evolution — an ape-like creature gradually evolving into "more advanced" species and culminating finally with Man — is wrong. Rather, *Homo sapiens* is the last remaining twig on what was a convoluted branch of the evolutionary tree. The various hominid species each occupied a niche, and each possessed various skills and attributes.

Our knowledge of earlier hominid species is sketchy, but we know much more about our closest relatives, the Neanderthals. (Our closest relatives still in existence are the great apes, with whom we share a common ancestor that lived some 5 million years ago.) It is instructive to remember the abilities and achievements of our sister species. Individual Neanderthals must have lived short, hard lives, but as a species they survived for a long time — much longer than mankind has been around; they inhabited a large area of Earth; they coped with severe swings in climate; in short, they successfully filled a biological niche. There is some evidence that Neanderthals buried their dead (though whether this practice was associated with the ritual accompanying modern human burials is doubtful).

There is also some slight evidence, from analyses of Neanderthal skulls, that they may have had the physical capacity for speech (though it seems more probable that they lacked the capacity to communicate in the way that we do). It is particularly interesting that they had a form of tool technology, called *Mousterian* (after the French cave of Le Moustier where such tools were first discovered). Mousterian tools are made of stone and take a variety of basic forms. The Mousterian craftsmen, then, were presumably able to hold several patterns of tool design in their minds and, combined with their deep appreciation of the properties of stone, produce quite beautifully constructed implements. The Neanderthals may not have matched the achievements of humans, but they were no mugs.[227]

However, during their period on Earth, Neanderthals demonstrated little in the way of creativity or innovation. If they created art, it has not survived; if they made music, their instruments have not survived. And their technology, though reasonably effective, was not subject to the sort of progress we have come to believe is inevitable. The late Mousterian tools were not significantly better than those of the early Mousterian. Neanderthals soon learned how to work stone, but then learned little else — not how to work bone or antler for tools, for example. So if we accept that Neanderthals were intelligent, then we have an example of an intelligent toolmaking species surviving for more than 100,000 years without making significant technological advance. They edged into extinction — for reasons not entirely clear — without inventing the ratchet much less the radio telescope. Perhaps this situation is mirrored on other worlds. Perhaps for some reason (lack of language, lack of a "creative spark," lack of hand–eye coordination, lack of whatever) alien species reach the level of toolmaking and then remain at that level. Perhaps the Galaxy abounds with species that are experts at handling wood or stone or bone, but that never develop further. We do not hear from ETCs because none of them have the required technology: in other words, *communicating* ETCs do not exist.

One weakness of this suggestion is that it requires *all* toolmaking species to develop in the same way. It is unconvincing in the same way that some of the "sociological" explanations fail to convince when they require all ETCs to *behave* in the same way. After all, even if hominid species in general have been poor technological innovators, one member of the hominid family is exceptionally innovative. One hominid species out of about 20 discovered the benefits of continual innovation; if that ratio is found elsewhere, the odds of finding ETCs would not seem so bad.

Before rejecting the suggestion completely, however, it is worth remarking that for much of our history we were not much better than the Neanderthals when it came to technological innovation. Only 40,000 or so years ago did our technology and art began to dazzle.[228] (The cave art of the

Cro-Magnons is truly dazzling. It is recognizably human and speaks to us across millennia. It is unlike anything appearing before that date.) Until this explosion of creativity, the two surviving hominid species appear to have been equally stagnant. Why the sudden change? There are several possible explanations. Perhaps the development of language triggered the creative explosion. Perhaps the explosion occurred much earlier, but artifacts prior to 40,000 years ago have not been preserved. Perhaps the humans of more than 40,000 years ago were anatomically modern, but did not possess modern brains. Or perhaps cultural knowledge accumulated slowly until, 40,000 years ago, it passed a critical threshold. We do not know. Perhaps whatever caused this explosion of creativity was a fluke, an accident. If it was, then we might expect the number of communicating ETCs to be small.

*   *   *

One last point. Inherent in the formulation of the Fermi paradox is the notion of an exponential growth in knowledge and technology. Perhaps most of us believe, consciously or otherwise, that early humans were in the "flat" part of the exponential curve: progress came slowly. Then, as time passed, progress fed upon itself and we end up today with computers obeying Moore's law. We extrapolate this exponential curve into the future and imagine our descendants having access to tremendously powerful technology; and, if ETCs are much in advance of us, we expect them to possess tremendously powerful technologies. But maybe this is wrong. In Nature, exponential curves never continue indefinitely. Perhaps the idea of technological progress continuing until a species can travel or at least communicate over interstellar distances is wrong.

This suggestion seems unduly pessimistic, at least to me. Even with our present technology, we can make a stab at communicating with the stars. Give humanity another 100 or 1000 years, and who knows what it will achieve.

SOLUTION 47: INTELLIGENCE AT THE HUMAN LEVEL IS RARE

Mind is the great lever of all things;
human thought is the process by which human ends are alternately answered.
Daniel Webster

When Fermi asked "where *is* everybody?", the "everybody" referred to *intelligent* extraterrestrial creatures. While the discovery of *any* life elsewhere would be profoundly important, it is *intelligent* life we search for. It is (presumably) only *intelligent* life that can travel between stars and with whom we can communicate, interact and learn from. But perhaps intelligence — the sort that can understand the laws of physics and construct radio telescopes — is rare in the Universe? As many as 50 billion species may have lived on Earth, but only one has the sort of intelligence required. Perhaps the development of intelligence is a fluke, so that the f_i term in the Drake equation is small.

There are many aspects to this question, but there is space here to address only two. First, what *is* intelligence? Second, how did it evolve?

What *Is* Intelligence?

In terms of SETI activities, we can reasonably define a species as intelligent if it can build a radio telescope. The problem with this definition is that mankind apparently became intelligent only about 50 years ago! So although in a practical sense it might be a good definition, it fails on philosophical grounds. There must be a better way of capturing the essence of intelligence.

A common approach is to define intelligence in terms of certain mental tasks that we find difficult, such as playing a decent game of chess or solving an algebraic equation. However, it is not much more difficult to write chess-playing programs or automatic equation-solvers than it is to perform the activities themselves. And this software manifestly does not possess intelligence. The sorts of activity humans and other animals do without thinking are much more difficult to program. No one has yet come close to programming a robot capable of navigating the world outside or of coping with the various challenges everyday life throws up. If finding food and avoiding danger are any measure of intelligence, then the average rodent is *much* more intelligent than the smartest robot. So if we want to appreciate what intelligence really means, and whether humans are unique in this regard, it might help if we understood something of animal intelligence. Unfortunately, if it is difficult to define intelligence in humans, it is even more difficult to define intelligence in animals.

*   *   *

Most people, if asked to rank non-marine animals in terms of intelligence, would probably rate man as the most intelligent animal, followed perhaps by apes, down through dogs and cats, down further to the likes of mice and rats, down even further to birds, and so on. It is a comfortable picture for the human ego: we are at the top of the tree of intelligence, our closest relatives are clever, our pets are quite bright, and the animals we do not particularly like are stupid. Implicit in this picture, though, is the notion of evolution as progress from a "less evolved" state (rats, say) to a "highly evolved" state (us), with intelligence being the scale against which one can measure progress. This is simply wrong.

In the first place, we have no reason for supposing intelligence (however defined) is the sole criterion by which we can rank animals. Why not instead use visual acuity, or speed, or strength? Indeed, why try to rank animals in this way at all? We should not view evolution as a ladder, with ourselves at the top and all other animals below us because they are not yet "evolved enough" to possess intelligence. Apes, birds, cats, dogs, mice and men are all equally "evolved," since we share a common ancestor that lived hundreds of millions of years ago. The various species have adapted to their environments in different ways; our species has certain characteristics that make it successful, but so has every other species on the planet. These species are all equally successful, since they have passed the critical test: they all have survived. If we want to assign different levels of intelligence to different animals, then we need a better gauge than our prejudices.

When biologists try to measure the intelligence of animals, they face an almost impossible task. Measuring the IQ of humans in a non-culturally biased way is difficult enough. But if tests on humans are biased, how can we possibly test the intelligence of different animal species? How can we factor out the differences in perceptual ability, manipulative ability, temperament, social behavior, motivation and all the other variations between species? Does a monkey fail to complete a maze because it is brainless or because it is bored? If a cat fails to press a lever that produces a food reward, should we conclude that the cat is stupid or is simply not hungry? Does a rat fail an intelligence test because it is dense, or because the test demanded visual discrimination (at which rats are poor) rather than discrimination between smells (at which rats excel)? These sorts of questions make it exceptionally difficult to be sure that we are testing an animal's cognitive ability.

Suppose we try to account for as many cross-species variables as we can think of in these cognitive tests. (For example, biologists might want to investigate how many list items an animal can remember, or whether an animal can recognize a face; either of these tasks might tell us something about cognitive processes in animals. The investigator would have to ensure the details of the test were different for different animals. The tests for pigeons and for chimpanzees would *have* to be different, if only to take into account their different physical abilities.) Suppose further that we define intelligence, *general* intelligence, to be a measure of how well animals score on such fundamental cognitive tests. Then a surprising fact emerges: most animals perform at about the same level! Of course there are some differences between species, but the differences are much smaller than one might expect. Chimpanzees can remember about seven items from a list at one time — but so can pigeons (so no more cracks about "bird brains"). Monkeys can quickly discern whether pile A contains more food treats than pile B — but so can cats. In fact, if intelligence is defined as the ability to perform these basic non-verbal tasks, then one can argue that to a first approximation all birds and mammals, *including mankind*, are about equally intelligent! This conclusion is still controversial; but if it turns out to be true we should not be surprised. After all, every species, including mankind, has to negotiate the same perilous world; we all have to eat and drink and find mates. The basic cognitive skills enabling animals to perform these tasks might well be common to all species.

On the other hand, one can equally take the opposite approach: maybe intelligence in animals consists precisely in all the factors we deliberately omit in cognitive tests. To use a computing analogy, we should not just consider the processor (the brain) but also the attached input and output devices (the senses and manipulative abilities of an animal). After all, a chimpanzee has hands which enable it to perform tasks that a cow simply cannot attempt. From this viewpoint there might be little general intelligence residing in the brain; rather, intelligence should be defined in terms of *specialized* intelligence — adaptations that enable particular species to succeed in their particular ecological niches. Support for this view is that the ability to learn (which is surely a large part of intelligence) seems to be specialized. Many animals can learn a particular task with ease but find it impossible to learn a logically equivalent task. It appears that an animal's learning ability depends upon the hard-wired behaviors already present in its brain. In this view, all animals are *differently* intelligent. It simply makes no sense to ask whether a bonobo is brighter than a homing pigeon: both creatures possess specialized intelligence enabling them to succeed in their particular environments.

These two seemingly opposite views of intelligence — that either *general* intelligence or *specialized* intelligence is the important factor — are perhaps merely two faces of the same coin. The lesson is that, cognitively, animals are both similar and yet different. In the case of mankind, much as we might like to think otherwise, our similarities with other animals are clear: we are simply not much better than many other animals at tasks that investigate fundamental, non-verbal cognition.

Nevertheless, it is impossible to deny the profound difference that exists between mankind and every other species. We may not be atop some evolutionary ladder of intelligence, but we *are* the only species capable of constructing abstract systems of thought. Only a member of our species can reflect upon his own thoughts and the thoughts of others. Only *Homo sapiens* is in the slightest bit interested in defining intelligence or wondering precisely what it means. Indeed, with an appropriate definition, one can quite reasonably exclude all other species and say that mankind alone is intelligent.

The Evolution of Intelligence

If for the moment we forget the details and simply use a "reasonable" working definition of human-level intelligence — that it involves a combination of factors including stereoscopic vision, symbolic language, tool use and so on — then one can ask the important question: how likely are other species to evolve a high level of intelligence?

Let us perform a thought experiment. Suppose 400 million years ago a meteorite had struck Earth, wiping out the ancestor of the vertebrate line but leaving untouched the ancestors of many other lines that are still alive today, such as the squid or the ant. Would any of those lines have given rise to intelligent species? Of course we cannot know for sure, because we live in a world in which vertebrates did not become extinct. But many evolutionary biologists think it improbable that human-level intelligence would have arisen from the squid-like ancestor or the ant-like ancestor. The reason is that evolution takes advantage of small, random mutations occurring in genetic DNA; if the change proves advantageous to an organism *in the here and now*, then the organism is competitively successful and the mutation propagates through the population. To repeat: evolution has no foresight. The mutation has to be of benefit *now*, not in the future, in order for the genes to spread. Now, there is no goal toward which evolution is working; much as we might like to think that high intelligence is the pinnacle of evolution, it simply is not. So, given this random process, the probability of producing the *same* complex adaptive feature from *different* evolutionary lines is tiny. The probability is small that the ancestor of the present-day squid could have given rise to a line that developed high intelligence.

What many SETI scientists pin their hopes on, however, is the phenomenon of *evolutionary convergence*. Sometimes different evolutionary lines arrive at the same solution to the only problem that matters — namely, keeping an organism alive long enough for it to pass on its genes. The classic example of evolutionary convergence is flight: birds, dinosaurs, fish, insects, mammals and reptiles all *independently* evolved the ability to fly. Another oft-cited example is the streamlining of marine creatures: species widely separated in evolutionary terms can nevertheless look similar. But these are convergences at low levels of complexity. It is not surprising that different creatures found that being airborne was a good way of escaping from predators, or that separate species discovered the benefits of cutting

221

quickly through water. So the relevance of these examples of convergence to the SETI debate is minor. Enthusiasts of SETI have always argued that a more convincing example of convergent evolution is the eye.

The eye is an incredibly complex and specialized piece of machinery. That it can evolve at all is really rather wonderful. Yet it seems to have evolved independently at least 40 times, and perhaps as many as 65 times. Furthermore, eyes employ at least a dozen fundamentally different designs. For example, the compound eye of the insect is totally different in design from the camera eye of the vertebrates; it seems that the eyes of insects and vertebrates must have evolved separately. Even eyes that appear superficially to be the same — for example, that of the squid and of man — on closer examination show differences in detail. And when you consider that the last common ancestor of squid and humans was probably a sponge-like creature that lived half a billion years ago ... well, it seems certain that the two types of eye evolved separately. That they look the same is a perfect example of convergent evolution. Or is it?

In 1993, Walter Gehring and Rebecca Quiring were studying the genetics of fruit flies.[229] They found a gene — called *eyeless* — that seemed to act as a master control gene for the formation of an eye in fruit flies. By suitable manipulation, they could "turn the gene on" in different places and have a fly sprout an ectopic eye on its wing or its leg or its antenna. *Eyeless* was not the gene "for" an eye — the way genes work is much more subtle — but it seemed, among other functions, to orchestrate the action of thousands of other genes that form an eye in the early development of an embryo.

It soon became clear that the fly *eyeless* gene was similar to a mouse gene called *small eye*. A mouse with a defective *small eye* gene develops shrunken eyes. Furthermore, the gene is similar to a human gene responsible for the condition Aniridia, sufferers of which can have defects of the iris, lens, cornea and retina. When geneticists made a detailed comparison it was discovered that the "eye genes" in these three quite different species — fruit fly, mouse and man — were essentially identical in two crucial locations.

Georg Halder and Patrick Callaerts decided to implant the mouse *small eye* gene into a fruit fly. The gene worked. It caused the fly to develop ectopic eyes — fruit fly eyes, not mouse eyes. The eyes were not wired to the brain, but they looked like normal insect compound eyes and they responded to light.

All of the phyla that scientists have studied carry some form of the *eyeless* gene. These findings cast doubt on the received wisdom that eyes are an example of convergent evolution, because if animals really did evolve the design of their eyes independently, then one would expect them also to have evolved their own genetic signaling system. There would be no reason why a mouse gene could control the development of a fly's eye — one would expect them to "speak different languages." Perhaps, then, the last common ancestor of phyla as diverse as vertebrates, cephalopods, arthropods and nemerteans already had an eye and a version of the *eyeless* gene. The jury is still out, but it seems increasingly likely that the eye evolved only once — and the different visual systems we see around us are the result of evolution playing variations on an existing theme.

If the eye arose only once, then what chance is there of something even more complex — high-level intelligence — arising independently from different evolutionary lines?

SOLUTION 48: LANGUAGE IS UNIQUE TO HUMANS

> ... *I learn'd the language of another world.*
> Lord Byron, *Manfred*, Act III, Scene 4

Ludwig Wittgenstein once famously remarked that "if a lion could talk, we would not understand him." It is easy to see the philosopher's reasoning: lions must perceive the world in ways quite alien to us. They possess drives and senses we simply do not share. On the other hand, the statement is all wrong. If a lion spoke English, then presumably English speakers *could* understand him — but the mind of that lion would no longer be a lion's mind. *The lion would no longer be a lion.* Humans talk; lions do not.[230]

Some people argue that humans are unique in being the only species in the history of Earth that has employed language. If language developed in only one species — just one out of the 50 billion species that have ever existed — then we might infer that the likelihood of language developing is small. Perhaps it developed in humans just through dumb luck — a chance assembly of several unlikely physical and cognitive adaptations. We are unique on Earth, and we may be unique in the whole Galaxy: perhaps humans are the only creatures that can talk. And since language opens up so many possibilities — so much of what we do individually and socially would not take place in the absence of language — creatures without language would surely be unable to build radio telescopes. No matter how intelligent those creatures might otherwise be, if they had no language, then we would not hear from them.[231]

Could this explain the Fermi paradox? Perhaps many planets are home to advanced life, but only here on Earth has a species learned to talk. At first glance it seems to be an outrageous suggestion, but it becomes more plausible upon closer perusal.

*   *   *

Noam Chomsky, one of the most profound thinkers of our age, has done more to elucidate the nature of human language than has anyone else.[232] Chomsky argues that language is *innate*. A child does not *learn* language; rather, language *grows* in the child's mind. In other words, a child is genetically programmed with a blueprint — a set of process rules and simple procedures that make the acquisition of language inevitable. All of us have a "language organ" — not something you can cut out with a knife, but rather a set of connections in the brain dedicated to language in the same way that parts of the brain are dedicated to vision. In this view, language acquisition happens to a child in much the same way that body hair suddenly sprouts on a pubescent teenager; it is part of growing up. Language is part of our genetic heritage.

Although Chomsky's ideas have been attacked by both adherents to the Standard Social Science Model (who argue that human practices within a social group are moulded by the culture of the group) and philosophers (who argue with Chomsky on several grounds), his theory seems to be the only way to explain several puzzles regarding language acquisition.

For example, language is an infinite system. If I were to speak this present sentence out loud, then there is an excellent chance that I would be the first person ever to utter these particular words in this particular order; it is a unique combination. One can construct an infinite number of sentences from a finite number of words. In order to cope with this infinite set, the brain *must* be following rules rather than accessing a store of responses. And when one considers what a child hears when its parents and siblings talk to it — just a sequence of sounds, including meaningless "uh's," "huh's" and "coochy-coo's" interspersing the poorly formed and incomplete sentences we all inevitably utter — it is remarkable that children develop and employ complex grammars so rapidly (all without the benefit of training, and often without feedback on the errors they make). Remarkable, that is, unless children are innately equipped with a *language acquisition device* (LAD) that lets them pluck the relevant syntactic patterns from the gobbledygook assaulting their ears. There is just one LAD, common to all humanity; there is not one device for Albanian, another for Basque, and yet another for Czech. Any child — so long as he or she receives sufficient stimuli to trigger the LAD at the correct age — can learn to speak any language. The stimulus need not even be auditory. If they are exposed to signing at the right age, hearing children of deaf parents can acquire sign language.

The operation of the human LAD may be similar to the innate visual acquisition device (VAD) of many animals. Scientists have performed experiments on kittens, blindfolding them immediately after birth. If the blindfold is removed any time before the first 8 weeks, the normal development of the kitten's visual system is resumed, and the adult cat will see normally. If the blindfold is kept on for longer than 8 weeks, the cat will suffer permanent visual impairment. It seems, therefore, that there is a critical period in which the VAD must receive external visual stimuli in order to establish the appropriate neuronal connections in *specific* pre-wired locations in the kitten's brain. If the connections are not established within this period, the chance of developing a fully functioning visual system is lost. Other parts of the brain cannot act as stand-ins for the visual system. The same effect has been observed to occur in those tragic cases in which linguistic input is withheld from children during the critical period up to puberty: their ability to speak grammatically is severely impaired. The existence of a critical period for language acquisition is not necessarily mysterious: it is presumably simply part of the same genetically controlled maturation process that causes our sucking reflex to disappear, our baby teeth to erupt, and all the other changes that occur to the human body. It makes evolutionary sense for the LAD to switch on early, as that way we have the maximum time to enjoy the considerable benefits of language. It also makes sense for the LAD to switch off when its job is done, since maintaining the device presumably incurs considerable costs in terms of energy requirements.

Although different languages differ in the specifics, there is a universality to language. And it is these universal *principles* that are innate. When a child develops language, then, the procedure follows an internal, predetermined course. A child who is acquiring Dutch will set the parameters of this predetermined system in one way; the child acquiring English will set the parameters in another way; and the child acquiring French will set the parameters of the system in yet another way. But the underlying principles are the same. To use a software analogy, language acquisition is rather like a macro with arguments — one argument for each language. (Vocabulary, of course, must be learned: if individual words were innate, then a neologism like "pulsar" would have to be assimilated into the gene pool before astronomers could use it! Cultural evolution would move at the same glacial pace as genetic evolution. Certain grammatical constructions must also be learned. For example, there is rule for forming the regular past tense of an English verb — namely, add *-ed* — but the past tense of irregular verbs must be learned on a case-by-case basis.)

In addition to evidence from linguistics and from the study of language acquisition in children, clinical evidence is at least consistent with the notion that language is innate. In some unfortunate patients, trauma or disease harms particular locations in the brain — locations that appear to be responsible for language processing. The effects can be distressing. For example, patients in which Wernicke's area is damaged find it difficult to comprehend the speech around them. More bizarrely, they suffer from Wernicke's aphasia: their speech is rapid, fluent, filled with grammatically correct phrases — yet their speech make no sense. They often substitute one word for another, and they coin new words; when asked to name objects, they give semantically related words or words that distort the sound of the correct word. Transcripts of their speech can make for disturbing reading — like reading the ramblings of a psychotic. On the other hand, patients with damage to Broca's area suffer from Broca's aphasia — speech that is slow, halting and ungrammatical. They can often comprehend the speech going on around them, or at least make informed guesses as to the meaning of speech, thanks to their prior knowledge of the world and the built-in redundancy of speech. (They can understand a sentence like "the cat chased the mouse" because they know cats chase mice.) Patients in which the *connection* between Wernicke's and Broca's areas is damaged suffer a form of aphasia that renders them incapable of repeating sentences. Even worse is the aphasia affecting patients in which Wernicke's and Broca's areas, and the connection between them, are undamaged but isolated from the rest of the cortex. The patients can repeat what they hear but have no understanding of what they are saying; they never initiate conversation. In yet other cases, damage to specific parts of the brain — often through stroke — causes remarkably specific language problems. Some aphasics can recognize colors but not name them; others cannot name food items, though they know what they like to eat; others cannot name items of clothing but have no trouble dressing themselves. At present, neuroscientists cannot map the brain and highlight different areas as handling different aspects of language. However, the evidence is that language is localized. And although localization itself does not mean language is innate, it does suggest we have a language organ.[233]

If we possess an innate language faculty, then the obvious question is: how did we come by such an intricate and complex organ? The answer is equally obvious: it evolved by natural selection of heritable variations. Unless we invoke the involvement of a creator, natural selection is the only known process that can generate such wonderful structures. If our language organ is the result of evolution, though, should we not see traces of it in the apes? After all, we are descendants of apes, aren't we? Well, no, we are not. Humans and apes are linked by a common ancestor that perhaps lived as long ago as 7 million years. It is entirely possible that our LAD evolved some time within the last 7 million years, so that it is not shared with the evolutionary branch leading to modern apes. Indeed, it has been suggested that the minds of early modern humans of about 100,000 years ago contained several separate "modules": a module for language, a module for technical intelligence, a module for social intelligence, a module for natural history, and so on. It may be that these isolated modules began to communicate only 50,000 years ago; and only then could people get together in groups and discuss, for example, the merits of a new tool design for use in hunting. (Perhaps it was only then that human consciousness, as we now understand it, developed. Only then did we become fully human.)

*   *   *

The followers of Chomsky would argue that language is specific to the human species. If you want to understand other animals, study what they do best; but it is pointless studying their language capabilities, since language is a human-specific ability. Pigs do not fly; neither do they talk.

But are we *sure* we are unique? What about chimpanzees, or dolphins, or dancing bees — do they not communicate in their own way? Perhaps they have innate language abilities too. One of the difficulties in contemplating these questions is our language: we seem compelled to anthropomorphize. Even when describing inanimate objects we anthropomorphize: genes are "selfish," the car is "acting funny," my chess program is "figuring out" the best move to make. There is of course nothing wrong with employing metaphor — assigning intentionality to inanimate objects enables us to convey the appropriate thought quickly — but sometimes we can forget that anthropomorphic statements do not necessarily describe what is *really* happening. We have to be careful when describing an animal's actions in terms of our own conscious thoughts and motives. When we describe an animal as communicating some word or idea — effectively, when we say it is "talking" — we might be completely wrong.

Here is an example where our first interpretation of events may be wrong. Some types of ground squirrel living in open country suffer two main predators: hawks, relying on speed, attack from the air, while badgers, relying on stealth, attack from the ground. When a squirrel spots a predator it chooses (there is an anthropomorphic usage!) from one of two defensive strategies. If it spots a badger, then the squirrel retreats to the opening of its burrow and maintains an erect posture. A badger, seeing that posture, knows the squirrel has spotted it and thus an attack would be a waste of time and energy. If a squirrel spots a hawk, then it runs like hell for the nearest cover. Squirrels also emit two different alarm sounds. If they spot a badger, then they make a rough chattering sound; if they spot a hawk, then they emit a high-pitched whistle. Other squirrels in the vicinity react when they hear the sounds, retreating to their burrows when they hear the badger alarm or running for cover when they hear the hawk alarm. Our natural inclination is to think squirrels are communicating with each other; that they are saying in effect: "Careful, now, there's a badger around; better head for home" or "Oh-oh, hawk; get *out* of here!" But are they?

As its actions upon spotting a predator clearly show, any individual squirrel is interested in saving its own skin. Indeed, evolutionary theory tells us that this must be the case: a squirrel could not care less about the fate of other squirrels. But if the squirrel alarm calls carry semantic information — if they are calling out "brock!" or "hawk!" in squirrelese — we meet a paradox. Selection will favor those squirrels who keep quiet, sneak off silently, and let the other suckers get eaten; being a non-caller in a group of callers is selectively advantageous, and the squirrel gets to pass on its genes. Soon, though, you end up with a community of silent squirrels; where does the instinct to cry out arise? The behavior of the squirrels makes sense only if their calls do *not* convey semantic information. Consider the squirrel's "hawk alarm." First, it is a high-pitched whistle — which, as experiments have shown, hawks find difficult to locate. So the squirrel is revealing nothing to the hawk. Second, being the only one to run for cover makes a squirrel conspicuous; it is much better to be one of a group of squirrels that are scrambling around, because the chances of being singled out by the hawk are reduced. Similarly, squirrels that run for cover when they hear a high-pitched whistle are less likely to be eaten by a hawk than squirrels that stand their ground. So selection will tend to favor squirrels that cry out when they see a hawk, and also those that run for cover when they hear a high-pitched whistle. When *humans* look at the situation they interpret it as squirrels sharing information. But that is not what is happening. The behavior is simply a trait that is passed on through the generations because it is effective. The squirrels do not even have to be aware of one another for this sort of behavior to evolve. No words; no language; just the forces of evolution. A similar analysis can be applied to the famous case of the vervet monkeys, which have "alarm calls" for eagles, leopards and pythons.

But what about chimpanzees like Washoe and bonobos like Kanzi, which have been taught American Sign Language (ASL)? Surely the achievements of these creatures proves that some animals have the capacity for language? Even here we must be careful. The team of scientists who trained Washoe for three years claimed at the end of the program that the chimpanzee could use 68 signs, and even string some of them together in two- and three-word sentences. Herbert Terrace, a scientist who was entranced with the idea of communicating with another species, sought to replicate the experiments. He raised the chimpanzee Nim Chimpsky (the reason for the name should be obvious!) in a highly social setting and taught it a set of ASL signs. Terrace videotaped the signing sessions and, after analyzing the data from these sessions, completed most of a book describing Nim's success in acquiring sign language. Then, when he replayed the tapes in slow motion for a final analysis, Terrace made a discovery:

nearly all of Nim's signs were prompted by its human teachers. Furthermore, the chimp's signs were often imitative of what its teachers had just signed. Nim was never spontaneous with its signing; the signs were made to obtain rewards from its teachers (and even then, it resorted to signs only after more direct methods of obtaining a reward had failed). In short, Nim did not display anything like full language. When scientists scrutinized the publicly available tapes of Washoe it was clear the same thing had happened: the chimp was imitating signs that a trainer had just made. Perhaps the strongest criticism of the Washoe experiment came from the one native ASL signer on the team. He recalled how the scientists would log as a sign every vague movement Washoe made, even though the gesture might resemble no valid ASL sign. The scientists' conclusion was a case of wishful thinking. In a similar case with a gorilla called Koko, its trainer explained away Koko's many mistakes by calling them metaphors and mischievous lies. If you take that approach to data analysis, you can find anything. Even in the case of Kanzi, an animal that undoubtedly displays impressive cognitive abilities, great care must be taken not to over-interpret what it does. No matter how generous we are, we simply cannot argue that Kanzi uses language in anything like the way humans use it.

Using a system of rewards to train captive chimpanzees is one thing, but what chimps do in the wild is something else. There is absolutely no indication that chimps — or indeed any other creatures — use language spontaneously. Many other pieces of evidence suggest that animals do not possess symbolic language. For example, in one recent experiment scientists released a dolphin into one end of a pool containing apparatus that (once the dolphin had figured out how it operated) released food. The investigators timed how long it took the dolphin to understand how the apparatus worked, then transferred the dolphin to the other end of the pool. A barrier prevented the dolphin from swimming back to the apparatus, but it could still see the apparatus and could send signals through the water. The scientists released a second dolphin into the pool near the apparatus. On average, the second dolphin took the same time to operate the apparatus as did the first dolphin. We can conclude from this that the first dolphin was unable to tell the second dolphin how the apparatus worked. Dolphins lack an abstract language. A similar experiment with chimpanzees had the same results: the chimps could not communicate their knowledge.

As a final piece of evidence that our relatives lack an innate capacity for language, consider what happens when scientists excise the areas of a monkey's brain corresponding to Broca's and Wernikce's areas in humans: the monkey's ability to produce or respond to vocal calls is unaffected.

Although the suggestion that only humans possess symbolic language may be controversial, many people (myself included, for what it is worth) think it is self-evidently the case. Even if we *can* train certain animals to use words, no animals come close to using language in the abstract, spontaneous, playful, creative way that humans use language. It seems silly to deny the fact. It also seems arrogant and anthropocentric to measure the abilities of animals in terms of *our* capabilities. Birds can perform feats of navigation that no human can match without aids. Some marine animals can, unlike humans, sense electric currents. Dogs can hear sounds beyond our perception and smell scents to which our noses are dead. Bats use an incredible system of echolocation. Horses have been known to pick up on cues that humans miss completely. And so on. Every species has abilities, forged by evolution, which enable them to scrape a living in a world that cares not whether they survive. This diversity is wonderful, and should be celebrated. Defining other species in terms of how well or how badly they use *human* traits is to demean those species.

*   *   *

Articulate speech is vitally important to the success of our species. Perhaps it is impossible for any species to develop the ability to travel or communicate over interstellar distances if they lack some equally sophisticated method of communication. And yet, in the case of the evolution of human speech, we seem forced to conclude that articulate speech is the result of a series of chance environmental changes and evolutionary responses; it was just good luck. Consider, for example, what happened to the bodies of our ancestors: a restructuring of the human diaphragm, larynx, lips, nasal passages, oral cavity and tongue, all of which were vital for articulate speech to develop, but none of which occurred *in order* for speech to develop. The changes to these organs were initially completely unrelated to the capacity for speech; they were small changes that brought immediate selective benefits. At least one of the changes — the positioning of the larynx deep in the throat — seems bizarre. Having a larynx low in the throat provides the tongue with enough room to move and produce a large number of vowel sounds, but any food and drink we swallow has to pass over the trachea: choking to death becomes a distinct possibility. If the tape of life were replayed, perhaps humans would not develop language. The benefits are great, but so are the costs.

On Earth, of the 50 billion species that have existed, only humans possess language. Language enables us not only to think, but to think about the thoughts we have. It enables us to reflect upon our thoughts, to try out new patterns of thought, and to record our thoughts. Language is what makes us human. If we ever visit other worlds, perhaps we will find billions of other species — each well adapted to its particular niche, but none of them with the single adaptive trait we are searching for: language.

SOLUTION 49: SCIENCE IS NOT INEVITABLE

For science is like virtue, it's own exceedingly great reward.
Charles Kingsley, *Health and Education*

If an ETC is to communicate with us, presumably it will need to possess a high level of scientific knowledge. For only through science will it understand how to build a radio telescope (or some other device to enable interstellar communication). But even if an intelligent extraterrestrial species *does* learn to make tools, *does* develop technology, and *does* acquire language, will it then inevitably develop the methods of natural science?

Earlier we looked at a solution to the Fermi paradox that suggested that ETCs might develop a *different* science or mathematics. Here the suggestion is slightly different: perhaps there is only one approach to science, but so far only humans have found it. Perhaps the Galaxy is swarming with species more intelligent than us — creatures excelling in the arts and philosophy — but who lack the techniques of science. So we do not hear from these species because they cannot make themselves heard over interstellar distances. "They" — meaning, as always, intelligent *communicating* civilizations — do not exist.

Those who offer this as a resolution of the paradox — and it is implicit in thousands of SF stories — presumably take their cue from the historical development of natural science on Earth. Many civilizations developed mathematics and medicine, but the origins of natural science were much more restricted. Consider, for example, the Aborigines. Recent findings indicate that the Aborigines may have arrived in Australia as far back as 50,000 years ago — a landmark achievement in human history that is sadly underestimated. The culture of Australia's indigenous peoples is perhaps the oldest continuously maintained culture in the world; their stories and belief systems are the most ancient on Earth. They have lived in a wide range of environments with great success for an unimaginable length of time. Yet in all that time they never developed the techniques of modern science. It seems that science is not inevitable. The dawn of modern science only began about 2500 years ago with the Greeks; but, despite possessing some of the most brilliant scientists of all time, Hellenistic science was limited. It was shackled by a pervading intellectual snobbery that valued contemplation over experiment. It took almost 2000 years for science as we now understand it to really get underway, with scientists like Galileo and in particular Newton pioneering a quantitative approach to scientific reasoning. Why did it take so long for the seeds planted by the Greeks to flower into our modern scientific endeavors? And although science is now a global activity, why did the flowering take place in such a restricted geographical area?

After the demise of the ancient Greek civilization, many other civilizations developed sophisticated technologies and systems of mathematics. The Arabic civilizations in North Africa and the Middle East possessed some excellent mathematicians (much of our knowledge of Greek astronomy was preserved by them). The civilizations of South America had architects that built fantastic structures. The Chinese civilization was for many hundreds of years the most advanced on Earth. Yet none of them — nor any of the others civilizations around the world — developed the methods of modern science, and none of them developed the scientific approach to the study of Nature, which has proven to be so powerful. Why?

It may be that cultural factors played a role. For example, some authors believe the prevailing philosophy of the Chinese civilization encouraged a "holistic" view of the world, so it was more difficult for them to take a Western "analytic" approach to science. Newton was ready to consider a system in isolation from the rest of the Universe and apply his techniques to that idealized, simplified system. Had he attempted to provide a complete description of Nature in all its messy holistic complexity, he would surely not have succeeded. And in 1709, while the world was still absorbing the impact of Newton's great scientific books, the spark that ignited the industrial revolution — Abraham Darby's use of coke rather than charcoal for smelting iron — took place in Ironbridge, England. At the same time in China, a centuries-old iron works was in the process of being closed. The Chinese thought they had no further need of it.

Some authors, then, argue that the development of science is far from inevitable. There is a variety of reasons — luck, environmental hindrances, cultural factors, philosophical inclination — why ETCs might not hit on the techniques of science.

Yet it is hard to accept this as a plausible explanation of the Fermi paradox. Yes, it took almost 2000 years between the emergence of Hellenistic science and the rise of modern science. This is a long time on the human scale, but remember as always that this is not the correct timescale with which to consider these questions. In the Universal Year, 2000 years corresponds to 5 seconds. On a cosmic timescale it matters not at all that natural science was developed by a Western European civilization rather than the Incas, Ottomans or Chinese. Had it taken mankind a further 2000 years (or 20,000 years) to invent science, it would make little difference as far as the Fermi paradox is concerned.[234] The scientific method had to be invented only once: it was so effective that it spread quickly, and is now the common heritage of our species. Should we not expect the same to be true for ETCs?

6

Conclusion

After criticizing 49 resolutions of the Fermi paradox, it is only fair that I present my own. It is not an original suggestion, but it sums up what I feel the paradox may be telling us about our Universe.

David Brin, in his superb 1983 analysis of the "great silence,"[235] wrote that "few important subjects are so data-poor, so subject to unwarranted and biased extrapolations — and so caught up in mankind's ultimate destiny — as is this one." Almost two decades later, little has changed.

The subject is *still* important. What could be more so? Either we are alone, or else we share the Universe with creatures with whom we might one day communicate. Either way, it is a staggering thought.

The subject is *still* data-poor. To be sure, there have been advances in specific areas. Advances in computing and astronomical technology have made possible the development of powerful SETI programs, and we now know much more about the formation of planetary systems and of the evolution of life on Earth (although in both these cases, as is usual in science, new discoveries seem to create an expanding shell of ignorance). Nevertheless, we have barely begun to find answers to many of the deep questions.

And the subject is *still* liable to unwarranted, biased extrapolations. Given the profound importance of the subject, though, should our lack of hard data force us to remain silent? Surely the best we can do under the circumstances is to be frank about our biases and open about our extrapolations. At least then a debate can take place, even if for the moment such debate will generate more heat than light.

Chapter 6

SOLUTION 50:   THE FERMI PARADOX RESOLVED ...

When facts are few, speculations are most likely to represent individual psychology.
Carl Gustav Jung

The paradox resolved? Well, not really. The topic remains so intangible that honest people can reach quite opposite conclusions. The reader is free to choose one of the solutions presented earlier, or to originate his or her own. Here, though, I present the solution that makes most sense to me.

*   *   *

There is just one gleaming, hard fact in the whole debate: we have not been visited by ETCs, nor have we heard from them. So far, the Universe remains silent to us. Those who would deny this fact of course have a ready solution to the Fermi paradox (and presumably stopped reading this book after the first few pages). The job for the rest of us is to interpret this lone fact.

As the above quotation suggests, with just one piece of evidence to play with, our biases will come to the fore. My own biases, such as I can identify them, include optimism about our future. I like to think our scientific knowledge will continue to expand and our technology to improve; I like to think mankind will one day reach the stars — first by sending messages and then later, perhaps, by sending ships. I like to think something akin to the Galaxy-spanning civilization described by Asimov in his classic *Foundation* stories might one day come to pass. But these biases collide with the Fermi paradox: if *we* are going to colonize the Galaxy, why have *they* not already done so? They have had the means, the motive and the opportunity to establish colonies, yet they appear not to have done so. Why?

Of the suggestions discussed in Chapter 4, only Solutions 16, 17 and 20 strike me as plausible resolutions of the paradox; I suspect that most SETI scientists would agree that some combination of these ideas is likely to be correct. (Strictly, these are solutions to the "great silence" question: why do we not *hear* from ETCs? To explain why ETCs have not visited us, or why we see no evidence of their existence, we must take further suggestions into account — that interstellar travel is impossible, for example.) But the only position that is consistent with the observed absence of extraterrestrials and that at the same time supports my prejudices — the only resolution of the Fermi paradox that makes sense to me — is that we are alone.

*   *   *

If you look up at the sky on a clear moonless night and gaze with the naked eye at the myriads of stars and the vastness of space, it is difficult to believe we might be alone. We are too small and the Universe is too big for this to make sense. But appearances can be deceptive: even under ideal observing conditions you are unlikely to see more than about 3000 stars, and few of those would provide conditions hospitable to our form of life. The gut reaction we perhaps all feel when we look at the night sky — that there *must* be intelligent life somewhere out there — is not a good guide. We have to be guided by reason, not gut reaction, when discussing this matter. Well . . . reason tells us there are a few hundred billion stars in our Galaxy alone, and perhaps a hundred billion galaxies in the Universe. Just one sentient species when there is such an immense number of places life might get started? Come *on* . . . surely I cannot be serious?

When discussing some of the different types of paradox, I noted Rapoport's observation that the shock of a paradox may compel us to discard an old (perhaps comfortable) conceptual framework. I believe the Fermi paradox provides a shock that forces us to examine the widespread notion that the vast number of planets in existence is sufficient to guarantee the existence of extraterrestrial intelligent life. In fact, we need not be too surprised. The Drake equation is a product of several terms. If one of those terms is zero, then the product of the Drake equation will be zero; if several of the terms are small, then the product of the Drake equation will be very small. We will be alone.

If one factor in the Drake equation is close to zero, then we can reasonably identify that factor as being *the* solution to the Fermi paradox. For example, as we saw in Solution 30, some scientists argue that the emergence of life was an almost miraculous fluke, a 1-in-10^{100} event (a number that dwarfs the number of planets in the Universe, and when expressed as a probability becomes, for practical purposes, indistinguishable from zero). Other scientists argue, perhaps more convincingly, that the improbability of the prokaryote–eukaryote transition (Solution 44) explains the paradox. Rather than there being a *single* solution to the paradox, however, I suspect there is a combination of factors — a product of various solutions we have discussed in this book — resulting in the uniqueness of mankind.

It is usual at this point to pick some numbers favorable to one's position, plug them into the Drake equation, and then put forward the required result. I would prefer to present here a more pictorial approach.

*   *   *

When I was a schoolboy, I was fascinated by the Sieve of Eratosthenes.[236] Eratosthenes was a Greek astronomer and mathematician, famed for being head of the Library at Alexandria and for being the first to provide an accurate measurement of Earth's circumference. He also developed a technique — his "sieve" — for finding all prime numbers less than some given number N. Primes — numbers evenly divisible only by themselves and 1 — are extremely important in mathematics; they are like atoms, from which we can compose all other numbers through multiplication. If you are given a number at random, it can be difficult to know whether it is composite or prime. The Sieve of Eratosthenes is a technique for sifting out the composite numbers and leaving only the prime numbers standing.

Suppose you are a Greek mathematician who wants to find all primes less than or equal to 100. First, you take a sheet of papyrus and write down the numbers from 1 to 100. The number 1 is special, so ignore it. The number 2 is prime so leave it; but go through the list and cross out all its multiples: 4, 6, 8, ... 100. Repeat the process, using the next smallest remaining number, 3; leave it because it is prime, but cross out its multiples all the way up to 99. Continue until you reach the end of the list. Remarkably quickly, you find all the numbers up to 100 have been deleted — except for the 25 prime numbers, which are still standing. Even for a computer, the Sieve of Eratosthenes is the quickest way of finding all primes less than about 10^8.

$\boxed{1}$	2	3	4_2	5	6_2	7	8_2	9_3	10_2
11	12_2	13	14_2	15_3	16_2	17	18_2	19	20_2
21_3	22_2	23	24_2	25_5	26_2	27_3	28_2	29	30_2
31	32_2	33_3	34_2	35_5	36_2	37	38_2	39_3	40_2
41	42_2	43	44_2	45_3	46_2	47	48_2	49_7	50_2
51_3	52_2	53	54_2	55_5	56_2	57_3	58_2	59	60_2
61	62_2	63_3	64_2	65_5	66_2	67	68_2	69_3	70_2
71	72_2	73	74_2	75_3	76_2	77_7	78_2	79	80_2
81_3	82_2	83	84_2	85_5	86_2	87_3	88_2	89	90_2
91_7	92_2	93_3	94_2	95_5	96_2	97	98_2	99_3	100_2

FIGURE 72   *This figure shows what happens when you apply the Sieve of Eratosthenes to a grid of numbers up to 100. The bold numbers are primes, which are left untouched by the procedure. The gray boxes are composite numbers — those that can be created by multiplying two or more prime numbers. The subscript on a composite number indicates the smallest divisor of the number — the first prime that sifted out the number. The number 1 is special, and is not considered prime.*

As a schoolboy, I was intrigued by the way the Sieve caught more and more of the large numbers. The technique was inexorable: on large grids I found myself chopping down number after number. Since the distribution of primes thins out quickly the higher you count, there are long stretches where all the numbers have been crossed out — numbers that have failed to make it through the Sieve.

I picture something similar happening with the Fermi paradox. Imagine writing down a grid of numbers, from 1 to 1000,000,000,000, with each number representing an individual planet in the Galaxy. (I arrive at this number by multiplying the number of stars in the Galaxy, which is about 10^{11}, with an assumed average of 10 planets per star. In fact, the number of stars is probably greater than this, with some estimates suggesting that our Galaxy contains as many as 400 billion stars. On the other hand, the average number of planets per star is likely to be less than 10. So although a figure of 10^{12} planets is a rough guess, it may not be too wrong — and anyway, this hardly matters when all the other numbers in the problem are so vague.) We assign Earth the number 1, since the Earth is special: it is the only planet on which we *know* intelligent life exists. Now start applying a sieve — let us call it the Sieve of Fermi. (The process I describe here is not meant to be the *only* way of working the numbers. You may prefer different numerical values for the quantities I describe, but the process shows why we should not be surprised if we discover that we are alone.)

*   *   *

Step 1   In Solution 36 we briefly discussed the notion of a galactic habitable zone (GHZ) in which a star must reside before it can give rise to a viable planetary system. A recent suggestion is that the GHZ contains only 20% of the stars in the Galaxy. So cross out those numbers corresponding to planets not orbiting a star in the GHZ: with 10 planets per star, 2×10^{10} planets remain. Now make a second application of the Sieve.

Step 2   The bright O and B stars die too quickly for life to evolve around them; the dull K and M stars are too miserly with their energy for life to prosper. For life *as we know it*, we need consider only stars like the Sun. (As I stressed in earlier sections, this assumption may be an expression of chauvinism — or a failure of scientific imagination. But I think it is the best assumption we can make at this time.) Only about 5% of stars in our Galaxy are like the Sun; cross out numbers corresponding to planets not orbiting a Sun-like star, and 10^8 planets remain.

Step 3   Life as we know it requires a terrestrial planet to remain in the continuously habitable zone (CHZ) for billions of years. We discussed the narrow width of the CHZ in Solution 36. We also discussed some factors that may cause Earth-like planets to be more rare than we might suppose, such as the migration of Jupiters to the inner parts of a planetary system (Solution 37) and the possible scarcity of rocky planets (Solution 35). My own guess is that only 1% of planets will be both suitable for life and remain in a CHZ for billions of years. You may think a different figure is in order here (and one could argue for higher or lower figures), but 1% seems reasonable to me. So cross out numbers corresponding to planets that do not remain in a CHZ: 10^6 planets remain.

Step 4   Of the million planets that orbit in the CHZ of a Sun-like star that is itself in the GHZ, how many are home to life? If you believe the genesis of life is exceptionally rare (Solution 30), then the answer is: none. If you believe a special set of circumstances is required, such as life originating on a planet like Mars and then being transported via impact ejecta to an Earth-like planet (Solution 43), then the answer is: not many. I prefer to believe that life is a probable occurrence: that if conditions are suitable, then there is a good chance of cells evolving. Let us say that the chance is 0.5. Cross out numbers corresponding to planets on which life does not arise, and 5×10^5 planets remain. Half a million planets with life!

Step 5   The Universe is a dangerous place. We saw how destruction can come from the depths of space (Solution 39) and from closer to home (Solution 40). We also discussed how the rate of planetary disaster may be significant (Solution 38). On many planets, life may be snuffed out — or at least prevented from evolving into complex life-forms — by some disaster. My guess is that as many as 20% of planets may suffer such a fate. (This *is* just a guess, and it may be an overestimate.) So cross out numbers corresponding to planets on which disaster strikes: 10^5 planets remain.

Step 6   We saw how Earth's system of plate tectonics was important in the development of life (Solution 41) and also how the Moon plays a role (Solution 42). If both these factors are *necessary* for the evolution of complex life, then the number of planets with the sentient species we are searching for may be small. However, although I believe these phenomena *are* important in some ways, I have no feel for the numbers involved. So I will ignore these factors, and at this stage of the sifting process all the planets make it through: 10^5 planets still remain.

Step 7   Cross out numbers corresponding to planets where life never evolves beyond the prokaryotic grade (Solution 44). The development of the modern eukaryotic cell took aeons on Earth, which perhaps indicates that this step is far from inevitable. No one knows what fraction of planets with prokaryotes will go on to host complex multicellular life-forms; my own estimate of one in ten may be *very* generous. We are left with 10^4 numbers — ten thousand planets possessing complex multicellular life. Does that mean the Galaxy contains ten thousand ETCs? Unfortunately not, because we must make several further applications of the Sieve before we arrive at the number of species *with whom we can communicate*. Let us combine all these into one last pass through the sifting process.

Step 8   Cross out numbers corresponding to planets on which advanced life-forms do not develop tool use and the ability to continuously improve their technology (Solutions 45 and 46). Cross out numbers corresponding to planets on which advanced life-forms do not develop the type of abstract high-level intelligence we are familiar with (Solution 47). Finally, and to my mind crucially, cross out numbers corresponding to planets on which advanced life-forms do not develop complex, grammatical language (Solution 48). How many planets remain? Of course, no one knows; it is impossible to assign accurate probabilities to these matters. My feeling is that many of these developments were far from inevitable. The feeling arises because, of the 50 billion speciation events in the history of our planet, only one led to language — and language is the key that enabled all our other achievements to take place. My own guess, then, is that none of the planets make it through this final sifting process.

After applying the Sieve of Fermi I believe that *all* grid numbers will be crossed out, except the number 1. Only Earth remains. We are alone.

*   *   *

I believe that the Fermi paradox tells us mankind is the only sapient, sentient species in the Galaxy. (We are probably also unique in our Local Group of galaxies, since many Local Group galaxies are unlikely to possess a GHZ. Perhaps we are even unique in the whole Universe — although the finite speed of light means ETCs could now exist in very distant galaxies without us yet being aware of them.) Yet the Galaxy need not be sterile. The picture I have is of a Galaxy in which simple life is not uncommon; complex, multicellular life is much rarer, but not vanishingly rare. There may be tens of thousands of exceptionally interesting biospheres out there in the Galaxy. But only *one* planet — Earth — has intelligent life-forms.

Such a picture is often criticized as violating the Principle of Mediocrity. The picture seems to suggest that Earth, and mankind, is special. Is this not the height of arrogance?

Paradoxically, at least to my mind, the expectation that other sentient species *must* be out there itself smacks of arrogance. Or rather, it achieves the tricky feat of being both self-important and self-effacing at the same time. At the core of this expectation is the belief that *human* adaptations, attributes such as creativity, and general intelligence, that we think important, are qualities to which other Earth organisms aspire and alien creatures may possess in even more abundance. Allow us a few more million years, so the logic seems to go, and we might evolve into the cognitively, technologically and spiritually superior beings that already exist out there. But the converse of this position is surely false. Give chimps another few million years, so the reasoning goes, and they too will be as intelligent and creative as us. But why should they be? Chimpanzees are good at being chimpanzees; dolphins are good at being dolphins; elephants are good at being elephants ... Rather than patronizing these species for not exhibiting *human* characteristics, we should respect them on their own terms for earning a living in a harsh world that cares not whether they live or die.

On the other hand it is undeniable that mankind *is* profoundly different from every other species on Earth. We alone have language, a high level of self-consciousness, and a moral sense. We *are* special. But surely our uniqueness could not have arisen by mere chance, by the blind and random groping of evolution, could it? Well, why not?

As Stephen Jay Gould pointed out in a delightful analogy, we can account for any growth in the complexity of living organisms through a drunkard's walk effect.[237] Imagine a drunk leaning against a wall. A few meters to his right is a gutter. If the drunk takes random equal-sized steps to his left or to his right, then he *inevitably* ends up in the gutter. No force propels him to his right; he moves randomly, and at any time he is as likely to move to his left as to his right. But the wall eventually stops his leftward motion; over time, there is only one direction in which to move. Eventually, completely by chance, the drunk stumbles into the gutter. The same effect can explain any advance we might observe in the complexity of organisms. At one end we have a wall of minimum complexity that organisms can possess and still be alive. This wall is where life began, and where most life on Earth remains. Over time, evolution tinkers with more advanced organisms; when life itself was young, that was the only available possibility — evolution could not try out simpler designs, because its path was blocked by the wall of minimum complexity. Some of the new designs worked, in the sense that the organisms were adapted well enough in their immediate environments to survive long enough to reproduce. And so evolution staggered on, like a blind drunk, tentatively producing organisms of greater complexity. After almost 4 billion years of random tinkering, we end up with the living world we see today. But there was nothing *inevitable* about the process; the purpose of evolution was not to produce us. Play the tape of history again, and there is no reason to suppose *Homo sapiens* — or any equivalent sentient species — would play any role at all.

Many eminent scientists argue that Mind is in some way predestined in this Universe. That far from being the outcome of chance, Mind is an inevitable outcome of deep laws of self-complexity. They argue that, over aeons, organisms will inevitably self-complexify and form a "ladder of progress": prokaryote to eukaryote to plants to animals to intelligent species like us. It is a comforting idea, but I know of no definite evidence in its favor, and I believe the silence of the Universe argues against it.

The famous French biologist Jacques Monod wrote that "evolution is chance caught on the wing." Even more evocatively, he wrote that "Man at last knows he is alone in the unfeeling immensity of the Universe, out of which he has emerged only by chance."[238] It is a melancholy thought. I can think of only one thing sadder: if the only animals with self-consciousness, the only species that can light up the Universe with acts of love and humor and compassion, were to extinguish themselves through acts of stupidity. If we survive, we have a Galaxy to explore and make our own. If we destroy ourselves, if we ruin Earth before we are ready to leave our home planet ... well, it could be a long, long time before a creature from another species looks up at its planet's night sky and asks: "Where *is* everybody?"

7

Notes and Further Reading

[¹ Pg 1]   The American author Isaac Asimov (1920–1992) was one of the 20th century's most prolific authors. He wrote on a vast number of topics — from the Bible to Shakespeare — but it was his science books, both fiction and non-fiction, that had the most impact on me.

The film *Forbidden Planet*, though now dated and containing some toe-curling dialog, in my opinion remains the best SF film of all time.

[² Pg 1]   The "pro-Fermi" article, by the American geologist and SF writer Stephen Lee Gillett (1953–), appeared in the August 1984 issue of *IASFM*. The rebuttal, by the American scientist and author Robert A. Freitas Jr., appeared in the September 1984 issue. A few years later, Gillett expanded upon his original article and pointed out that the "lemming paradox" is a non-paradox. If the Earth were empty except for lemmings, then lemmings *would* be everywhere; but the Earth is teeming with other living things, which out-compete lemmings and limit their spread. The correct conclusion to draw from the non-observation of lemmings is that Earth has

241

an abundance of living species that compete for resources (which we knew anyway, because we see life all around us). When we look into space, however, we see *nothing* that indicates the presence of life.

[3 Pg 5]   The reader who is unfamiliar with exponential notation needs to know only that it is a convenient method of handling very large and very small numbers. In this book I always use 10 as a base and so, in essence, the exponent counts the number of zeros following the 1. Multiplying numbers together using this notation is simple: just add the exponents. For example, $100 = 10 \times 10 = 10^2$ and $1000 = 10 \times 10 \times 10 = 10^3$. Division is just as simple: simply subtract one exponent from another. Thus $1000 \div 10 = 10^{3-1} = 10^2 = 100$. For numbers less than unity, the negative exponent counts the number of zeros following the decimal point. Thus $\frac{1}{100} = 0.01 = 10^{-2}$ and $\frac{1}{1000} = 0.001 = 10^{-3}$. Using exponential notation we can write, for example, 1 million as 10^6 and 1 billionth as 10^{-9}. This is useful in science, where we routinely deal with extremely large and extremely small numbers. Using exponential notation we can discuss the number of stars in the Universe (there are about 10^{22} stars) or the mass of an electron (its mass is about 10^{-36} kg) without resorting to unwieldy phrases like "a thousand billion billion" or "a trillion trillion trillionth."

CHAPTER 2   OF FERMI AND PARADOX

ENRICO FERMI

[4 Pg 8]   For details of Fermi's life I consulted two sources: a biography written by his wife [1]; and a readable account of Fermi's life in physics, written by Emilio Segré, a friend, student and collaborator of Fermi [2]. Segré (1905–1989) himself won the Nobel Prize for physics in 1959.

[5 Pg 8]   Luigi Puccianti (1875–1952), Fermi's teacher, was the director of the physics laboratory at the Scuola Normale Superiore in Pisa. Puccianti asked the young Fermi to teach him relativity. "You are a lucid thinker," Puccianti said, "and I can always understand what you explain."

[6 Pg 10]   The American physicist Arthur Holly Compton (1892–1962), who won a Nobel Prize for his work in subatomic physics, was in overall charge of the project that aimed to achieve the first self-sustaining nuclear reaction. When it was clear Fermi had attained the goal, Compton telephoned James Bryant Conant (1893–1978), the President of Harvard University. (Conant was a chemist but is now better remembered for his work as an educator.) The telephone call was cryptic: "Jim, you will be interested to know that the Italian navigator has just landed in the new world."

PARADOX

[7 Pg 12]   See [3] for an entertaining and readable book dealing with a variety of paradoxes. As well as the few I cover here, you can read about Russell's barber paradox, Newcomb's psychic paradox and many others (but not the Fermi paradox).

[8 Pg 12]   Beginning students of algebra often construct "proofs" of obviously untrue statements like $1 + 1 = 1$. Such "proofs" usually contain a step in which an equation is divided by zero; this is the source of the fallacy, since dividing by zero is inadmissible in arithmetic. If you divide by zero you can "prove" anything at all.

[9 Pg 12]   The Russian-born biomathematician Anatol Rapoport (1911–) is known for his work in a variety of fields, including the analysis of a famous mathematical paradox: the Prisoner's Dilemma. For a short, readable introduction to this paradox, see [4].

[10 Pg 13]   The liar paradox is only one of a number attributed to Eubulides (*fl.* 4th century BC). The sorites class of paradox based on "little by little" arguments is also often attributed to Eubulides. It is not known whether he invented all of these paradoxes nor, if he did, what his motives for doing so might have been. The famous Paulian version of the liar paradox appears in his letter to Titus, the first bishop of Crete (Titus 1:12–13).

[11 Pg 13]   The word *sorites* comes from the Greek word *soros*, meaning "heap," since it was first used in the type of reasoning described in the text. (In other words, one grain of sand does not make a heap; if one grain of sand does not make a heap, then neither do two grains; and so on *ad infinitum.*) See [5] for a comprehensive account of the sorites paradox.

[12 Pg 13]   The raven paradox was invented by the German-born philosopher Carl Gustav Hempel (1905–1997), one of the leaders of the logical positivist movement.

[13 Pg 14]   The paradox of the unexpected hanging was first noticed by the Swedish mathematician Lennart Ekbom when he heard the following wartime announcement by the Swedish Broadcasting Company: "A civil defense exercise will be held this week. In order to make sure that the civil defense units are properly prepared, no-one will know in advance on what day this exercise will take place." For more details on this paradox, see [6] by Martin Gardner. Although Gardner (1914–) is best known for his mathematics columns in *Scientific American*, he trained as a philosopher and has published scholarly articles on paradox.

Chapter 7

[^{14} Pg 14]   Zeno of Elea (*fl.* 450 BC) was a follower of Parmenides, a Greek philosopher who believed that the Universe consists of a single, undifferentiated substance. Our senses, of course, tell us that the Universe is anything *but* a "oneness"; we perceive many different substances. Zeno therefore set out to discredit the usefulness of human senses as a tool for discovering the nature of reality. He did so by presenting several paradoxes, in a book (sadly long since lost) on time, space and motion. Our senses lead us to believe in the existence of motion. But since Zeno "proved" that motion is logically impossible, our senses must be illusory — and we should therefore have no problem in accepting the rather strange beliefs of Parmenides. At least as important as the paradoxes themselves was the type of argument Zeno employed in them; Aristotle himself called Zeno the inventor of dialectical reasoning.

[^{15} Pg 14]   The resolution of Zeno's paradox came more than 2000 years after Zeno's death, when the Scottish mathematician James Gregory (1638–1675) developed techniques to handle convergent series. Gregory showed how an infinite series of numbers can have a finite sum. In the example given in the text, the infinite series $10 + 1 + 0.1 + \ldots$ has a sum of $11\frac{1}{9}$. In other words, Achilles overtakes the tortoise after $11\frac{1}{9}$ m.

[^{16} Pg 15]   Although the twin paradox involves Einstein's special theory of relativity, Einstein himself of course understood his own theory well enough not to present this phenomenon as a paradox! However, although Einstein was also one of the founders of quantum theory, he was less sure of his ground in this field. He and his co-workers — Boris Podolsky (1896–1966) and Nathan Rosen (1909–1995) — constructed a marvelously subtle argument (now called the EPR paradox) intended to prove that quantum physics is incomplete. Again, a full analysis shows there is no paradox — but at the expense of introducing a "spooky" (Einstein's own word) phenomenon called *entanglement*. The EPR result tells us that everything we have ever touched is invisibly tied to us by the weird rules of quantum theory. The best treatment of the EPR paradox can be found in [7]; see also [8]. The paradox was originally described in [9].

[^{17} Pg 15]   The dark-sky paradox was named after the German astronomer Heinrich Wilhelm Matthäus Olbers (1758–1840), but several other astronomers, including most notably Johann Kepler (1571–1630) and Edmond Halley (1656–1742), had considered the problem before Olbers published his analysis in 1826. See [10] for a thorough, elegantly written discussion of Olbers' paradox, including the early history of the question of why the sky is dark at night.

THE FERMI PARADOX

[18 Pg 17]   The first part of this section draws heavily on [11]. The author of that paper contacted Emil John Konopinski (1911–1990), Edward Teller (1908–) and Herbert Frank York (1921–), Fermi's luncheon companions on the day he asked his famous question, and requested them to record their recollections of the incident. During the early 1950s, the Americans Konopinski and York were both involved in theoretical work on the development of nuclear weapons, as was the Hungarian-born Teller (who has been described as "the father of the H-bomb"). All three of them would have enjoyed Fermi's input into discussions on nuclear physics.

[19 Pg 20]   The American astronomer Frank Donald Drake (1930–) was the first person in history to use a radio telescope to search for ETCs. A fascinating account of what led him to a life in astronomy, and of the prospects for finding ETI, can be found in [12].

[20 Pg 22]   Konstantin Eduardovich Tsiolkovsky (1857–1935) was born into a poor family in the city of Izhevsk in eastern Russia. From the age of nine he suffered almost total deafness following a streptococcus infection. Nevertheless, he educated himself and studied chemistry and physics. As early as 1898 he explained the need for liquid-fueled rockets for spaceflight, and in his 1920 SF novel *Beyond the Earth* he described how people would live in orbiting colonies. He promoted his ideas on extraterrestrial life in two essays entitled "There Are Also Planets Around Other Suns" (dated 1934) and "The Planets Are Occupied by Living Beings" (dated 1933).

[21 Pg 23]   For a description of Tsiolkovsky's philosophy and his anticipation of the Fermi paradox, see [13].

[22 Pg 23]   See [14].

[23 Pg 23]   Hart's classic paper generated interest in the paradox [15].

[24 Pg 23]   Lord Douglas of Barloch suggested [16] that the number of evolutionary steps leading from primitive life to intelligence was so large that the probability of it happening elsewhere was infinitesimal.

[25 Pg 24]   The American mathematical physicist Frank Jennings Tipler III (1947–) has written several popular articles on the use of probes to colonize the Galaxy. See, for example, [17].

[26 Pg 24]   Glen David Brin (1950–) trained as an astronomer, but is much better known as an award-winning SF writer. His article on the "great silence" [18] remains one of the clearest treatments of the subject. See also his popular article in [19], which gives a brief treatment of 24 possible solutions to the Fermi paradox.

[27 Pg 24]   See [20]. The updated second edition of this very readable book is easier to obtain than the first edition.

[28 Pg 24]   See [21] for a breezy account suggesting that the sheer number of stars in the Universe means there *must* be life elsewhere: give something enough of a chance to happen and eventually it will. However, many readers may find the arguments leading to this conclusion unconvincing.

[29 Pg 24]   See [22].

[30 Pg 24]   See [23].

[31 Pg 24]   Mention of economists reminds me of the Fermi paradox-like proof of the non-existence of time travelers [24]: if time travelers existed, then interest rates would not be positive! In fact, if people could travel back in time, then interest rates would have to be 0%, otherwise savers could use banks as bottomless ATM machines. Savers could simply travel back in time a few thousand years, deposit a few dollars, then return to the present; compound interest on even a small sum would guarantee riches.

[32 Pg 24]   A good example of the need for experiment is Tipler's argument that, in the distant future, we will all be resurrected in software by a God-like intelligence. [25] His argument rests on the Universe possessing certain cosmological properties; the latest observations seem to exclude these properties and thus Tipler's theory. We would not know this, however, unless astronomers had looked.

CHAPTER 3   THEY ARE HERE

SOLUTION 1:   THEY ARE HERE AND THEY CALL THEMSELVES HUNGARIANS

[33 Pg 28]   In [26], the author amusingly describes the Los Alamos "theory" that Hungarians are descended from Martians. The Hungarians at Los Alamos indeed formed an extraordinary grouping of talent. Edward Teller has already been mentioned. Leo Szilard (1898–1964) made contributions to molecular biology as well as nuclear physics — and also invented a novel type of home refrigerator; his co-inventor was Einstein! (See [27] for a good biography of Szilard.) Eugene Paul Wigner (1902–1995) was one of the foremost experts in quantum theory. John von Neumann (1903–1957) made immense contributions in a number of fields. Theodore von Kármán (1881–1963) was one of the world's foremost aeronautical engineers. All five were born in Budapest. Another physicist born in Budapest around the same time, although he did not work at Los Alamos, was Dennis Gabor (1900–1979). Gabor won the Nobel Prize for his invention of holography.

Such a grouping of talent is rare, but probably not unique. Other pockets of brilliance have occurred from time to time. For example, the 1979 Nobel Prize-winning particle theorists Sheldon Lee Glashow (1932–) and Steven Weinberg (1933–), who worked independently on electroweak unification, were in the same class at the Bronx High School of Science. Also in the class was Gerald Feinberg (1933–), who developed the idea of the tachyon. In addition to Glashow and Weinberg, the Bronx High School has produced three other Nobel Prize-winning physicists!

SOLUTION 2: THEY ARE HERE AND ARE MEDDLING IN HUMAN AFFAIRS

[34 Pg 29]   Ezekiel 1:4–28 contains a description of a wheel in the sky that some have chosen to interpret as a flying saucer. The interpretation of apocalyptic writings is notoriously difficult, but it is probably fair to say that the prophet Ezekiel was not describing a physical event. Depending upon one's outlook on these things, he could equally well have been describing a message from God, or he may have eaten some funny mushrooms.

[35 Pg 30]   Kenneth Arnold (1915–1984) wrote an account of his sighting in the 1952 privately published book *The Coming of the Saucers*.

[36 Pg 30]   The relatively early death of Edward J. Ruppelt (1922–1959), due to a heart attack, sadly but inevitably sparked more than a few conspiracy theories. A biography of Ruppelt, and a discussion of the 1950s UFO phenomenon from the point of view of "ufologists," is given in [28].

[37 Pg 31]   Many books have been written in support of the thesis that UFOs are alien spacecraft; sceptical approaches are much less common. One of the clearest sceptical essays on the UFO phenomenon is in [29].

[38 Pg 33]   The law of parsimony — the principle that entities are not to be multiplied beyond necessity — was probably first invoked by the French Dominican theologian Guillaume Durand de Saint-Pourçain (c.1270–1334). But William of Occam (1284–1347) applied the principle so frequently and so sharply that it became known as Occam's razor.

SOLUTION 3:   THEY WERE HERE AND LEFT
EVIDENCE OF THEIR PRESENCE

[39 Pg 34]   The Swiss Erich Anton von Däniken (1935–) wrote his most famous book, *Chariots of the Gods*, when he was working as a hotel manager. He followed it up with titles like *The Return of the Gods* and *The Gold of the Gods*. For an excellent and entertaining discussion of why these books are wrong-headed, see [30].

[40 Pg 35]   Five decades on, it seems strange to us that anyone could see a bridge on the Moon; but the Welsh astronomer Hugh Percy Wilkins (1896–1960) was a fine observer. He produced some excellent maps of the near side of the Moon, and was honored in 1961 by having a lunar crater with a 57-km diameter named after him.

[41 Pg 35]   For details regarding this argument, see [31].

[42 Pg 35]   The idea that a probe might observe Earth over a period of millennia is not so outlandish. Even with our present level of technology, the KEO project plans to put a passive satellite in orbit 1400 km above Earth's surface and have it stay in orbit for 50,000 years. The satellite, which will be launched in 2003, is a sort of time capsule; it will carry messages on a CD-ROM from people alive today (anyone can send a message to the project) and deliver them to whoever — or whatever — is then inhabiting Earth. It is an idea of French artist Jean-Marc Phillipe, who hopes to send a message to our descendants, just as the cave artists of Lascaux sent a message to us.

[43 Pg 36]   The Italian-French mathematician Joseph-Louis Lagrange (1736–1813) was one of the greatest mathematicians of the 18th century. Perhaps his most important astronomical investigations concerned calculations of the libration of the Moon and of the orbits of the planets.

[44 Pg 37]   An explanation of LDEs was given in [32]. The paper responded to the hypothesis [33] that LDEs were evidence of ETC probes at L4 or L5.

[45 Pg 37]   For an excellent account of observations of Mars, see [34].

[46 Pg 37]   The Italian astronomer Giovanni Virginio Schiaparelli (1835–1910), director of the observatory at the Brera Palace in Milan, made several important observations of meteors and comets before turning his attention to the planets in 1877. He was not the first to record channels on Mars; the first true map of Mars, published in 1830 by the German astronomers Wilhelm Beer (1797–1850) and Johann Heinrich von Mädler (1794–1874), contains at least one feature that seems to be a channel. Nevertheless, Schiaparelli so popularized the idea of *canali* that they became the defining theme of Mars. Perhaps the most famous of the stories that tapped into the public's subsequent fascination with the red planet was the superb 1898 novel *War of the Worlds* by English author Herbert George Wells (1866–1946).

[47 Pg 38]   Percival Lowell (1855–1916) came from a wealthy Boston family and did not take up astronomy in earnest until he was almost 40. He achieved quite a lot in science, despite his late start: he had the determination to initiate the search for a planet beyond Neptune, and the Lowell Observatory in Arizona is named after him. However, he will always be associated with his ideas concerning Mars.

[[48] Pg 39]   The Russian astrophysicist Josif Samuelevich Shklovsky (1916–1985) is best known for his explanation of continuum radiation from the Crab Nebula, but he also made important contributions in cosmic ray astronomy and on the distance scale for planetary nebulae. His popular book *Intelligent Life in the Universe*, which Carl Sagan translated and expanded, is a classic in the field.

The American astronomer Bevan P. Sharpless (1904–1950) worked at the US Naval Observatory; poor health hampered his work throughout his career. The fifth largest crater on Phobos is named after him.

[[49] Pg 39]   The Danish astronomer Heinrich Louis d'Arrest (1822–1875), director of the Copenhagen Observatory, mounted a thorough search for Martian moons in 1862. However, it was the American astronomer Asaph Hall (1829–1907) who discovered the moons, in 1877. The reason Hall found them and d'Arrest did not is simple: the Martian satellites are much closer to the planet than d'Arrest thought possible. Hall looked in the right place; d'Arrest did not. Thus, the suggestion by American biologist Frank Boyer Salisbury (1926–) that Phobos and Deimos were artificial satellites launched between 1862–77 is unnecessary.

[[50] Pg 41]   The Cydonian "face" was first pointed out in 1977 by American electrical engineer Vincent DiPietro. The view that the face is artificial has been championed most strongly by the American writer Richard C. Hoagland (1945–). See, for example, [35]. A more recent book in this vein is [36]. For a refreshingly sane article about the face, see [37].

[[51] Pg 42]   The Greek-American astronomer Michael Demetrius Papagiannis (1932–1998) was the first president of the International Astronomical Union's commission on bioastronomy. See [38].

[[52] Pg 42]   See [39].

[[53] Pg 42]   For an in-depth discussion of the possibility of detecting alien objects in the Solar System, see [40].

[[54] Pg 43]   Perhaps the first such serious paper was [41]. A later paper in a similar vein was [42].

SOLUTION 4:   THEY EXIST AND THEY ARE US — WE ARE ALL ALIENS!

[[55] Pg 44]   Anaxagoras (c. 500–428 BC), one of the greatest of Greek philosophers and the teacher of Socrates, spoke of the "seeds of life" from which spring all organisms. It was not until the 19th century, however, with work by Berzelius, Richter, Kelvin and Helmholtz, that the panspermia hypothesis took a modern form. Above all it was the work of the Swedish chemist Svante August Arrhenius (1859–1927), a man who helped lay the foundations of modern physical chemistry, which popularized the notion that life on Earth might have arrived from space [43].

[56 Pg 44]   The English astronomers Fred Hoyle (1915–2001) and Nalin Chandra Wickramasinghe (1939–) have made exceptional contributions to science, but they have also proposed several hypotheses that go against received wisdom. The Austrian-English-American physicist Thomas Gold (1920–) is another scientist who likes to propose unorthodox ideas. He jokingly proposed the "garbage" scenario for the origin of terrestrial life: ETCs landed here before life arose, dumped their rubbish, and the contamination from the garbage was the seed for life!

[57 Pg 45]   See [44]. The English-born biophysicist Francis Harry Compton Crick (1916–) gained fame for his discovery, along with the American biochemist James Dewey Watson (1928–), of the double-helix structure of DNA. He has continued to contribute to our understanding of the genetic code. The English-born biochemist Leslie Eleazer Orgel (1927–) has made major contributions to the study of life's origins. The Crick–Orgel idea of directed panspermia originated at the first conference on communication with extraterrestrial intelligence, organized in 1971 by Sagan and Kardashev, and held at the Byurakan Astrophysical Observatory in Armenia. Many of the luminaries in the field of SETI attended this conference.

SOLUTION 5:   THE ZOO SCENARIO

[58 Pg 46]   The American astronomer John Allen Ball (1935–) has written extensively on the Fermi paradox. For the zoo hypothesis, see [45].

[59 Pg 47]   Asimov's famous "humans-only" Galaxy was a reaction against Campbell's insistence that humans should always win out against aliens. Asimov thought that human civilization would be less advanced than any extraterrestrial civilizations we might encounter, and he could not bring himself to write stories in which Earth triumphed over superior alien technology. On the other hand, he wanted to sell stories to Campbell. He therefore removed the potential source of conflict, and his *Foundation* trilogy described a Galaxy containing only humans. If the Fermi paradox implies that we are alone, then perhaps an empire something like Asimov reluctantly described will come to pass.

[60 Pg 48]   The leaky embargo hypothesis [46] was proposed by James W. Deardorff (1928–), a retired atmospheric physicist. Although Deardorff has a scientific background, his leaky embargo hypothesis is unscientific. For a nice introduction to scientific method, which uses Deardorff's hypothesis as an example to be critiqued, see [47].

SOLUTION 6:   THE INTERDICT SCENARIO

[61 Pg 49]   The British writer Martyn J. Fogg (1960–) originally trained as a dentist. He is now one of the foremost authors on "speculative" engineering techniques, such as terraforming. His interdict hypothesis originally appeared in [48]. For a more popular account, see [49].

[62 Pg 49]   See [50] for a slightly dated, but still a good introduction to the subject. Asimov was an optimist and argued that half a million planets in our Galaxy are home to technological civilizations.

[63 Pg 49]   The notion of a "Codex Galactica" appears in [51]; note, however, that it is yet another idea that appeared in the pages of SF magazines before gaining respectability in the pages of a refereed journal.

SOLUTION 7:   THE PLANETARIUM SCENARIO

[64 Pg 51]   The British writer Stephen Baxter (1957–) is known for his "hard" science fiction. For details of his planetarium hypothesis, see [52].

[65 Pg 52]   Many examples exist of this paranoid trope in SF. The earliest such story of which I am aware is "The Earth-Owners" by Edmond Hamilton (1904–1977), which describes an Earth invaded by aliens in disguise; the aliens, of course, are busy manipulating us. Hamilton's story appeared in the August 1931 issue of *Weird Tales*. Historians of SF could perhaps point to even earlier examples. The Asimov story was "Ideas Die Hard" (*Galaxy*, October 1957). Weiner's "The News from D Street" appeared in the September 1986 issue of *IASFM*. The philosophical considerations underpinning the planetarium hypothesis are discussed in [53].

[66 Pg 54]   The Bekenstein bound is named after the Mexican-born US-Israeli physicist Jacob David Bekenstein (1947–), who introduced the concept in terms of the thermodynamics of black holes.

SOLUTION 8:   GOD EXISTS

[67 Pg 57]   See [54].

[68 Pg 59]   The Austrian-British philosopher Karl Raimund Popper (1902–1994) propounded the notion that scientific hypotheses must be falsifiable. The drive to falsify hypotheses is the essence of science. If an hypothesis cannot be tested and perhaps found to be false, then it is not a valid part of the process of science. Although his views about scientific progress have been attacked, they remain influential. Smolin's idea is certainly falsifiable, since it makes specific testable predictions; the novelty is that it must be tested by calculation rather than experiment.

[[69] Pg 59]   See [55].

[[70] Pg 59]   Asimov's haunting short story "The Last Question" tells how a pair of drunken technicians one night ask a supercomputer whether there is a way to reverse the increase of entropy and thereby halt the death of the Universe. The computer says there is insufficient data for a meaningful answer. The same question is asked of the computer six times over many different epochs. I will not spoil the story by telling you the computer's final answer!

CHAPTER 4   THEY EXIST BUT HAVE NOT YET COMMUNICATED

SOLUTION 9:   THE STARS ARE FAR AWAY

[[71] Pg 62]   A good place to learn about Voyagers 1 and 2 is the website given in [56]. See [57] for another NASA site with useful material on several of the advanced propulsion concepts discussed in this section.

[[72] Pg 62]   According to the theory of special relativity, massless objects such as photons always travel at light speed c, while objects with mass inevitably travel more slowly. Of course, it is possible to accelerate a slow-moving body to a faster speed by acting upon it with a force. Unfortunately for the prospects for space travel, special relativity tells us that the faster things move the more massive they become. At speeds close to c, the accelerating force tends to make the body more massive rather than make it move faster. The speed of light is a barrier that may not be reached by any object with mass — including space ships. For a good introduction to these concepts, see [58].

[[73] Pg 62]   See [59] for a discussion of astronomical distances.

[[74] Pg 63]   John Desmond Bernal (1901–1971), an Irish physicist, published the idea of a generation ship in a visionary book [60]. His book contains the following quote, which is relevant to any discussion of the Fermi paradox. "Once acclimatized to space living, it is unlikely that man will stop until he has roamed over and colonized most of the sidereal Universe, or that even this will be the end. Man will not ultimately be content to be parasitic on the stars but will invade them and organize them for his own purposes." For "man" read "ETC." So — where *are* they?

[[75] Pg 63]   The short novel *Universe*, by the American writer Robert Anson Heinlein (1907–1988), appeared in the May 1941 issue of *Astounding Science Fiction*. (It can be found more easily in [61].) The story is one of several SF classics penned by this author.

[⁷⁶ Pg 64] Time dilation is another of the unusual consequences of special relativity. Just as moving objects increase in mass, so moving clocks go slow. The faster a clock moves relative to an observer here on Earth, say, the slower that clock seems to tick compared to a clock carried by the Earthbound observer.

[⁷⁷ Pg 64] This possibility was dramatized by the American writer Poul William Anderson (1926–2001) in his novel *Tau Zero*. The novel tells the story of a ramjet that accelerates to speeds so close to c that circumnavigation of the Universe becomes possible. See [62].

[⁷⁸ Pg 64] For an interesting discussion of the problems inherent in navigating to a particular star, see [63].

[⁷⁹ Pg 65] As well as conceiving the idea of an antimatter rocket, the Austrian scientist Eugen Sänger (1905–1964) pioneered several practical ideas in rocketry. For a superb introduction to many different proposals for interstellar travel, see [64]. A more recent source is [65].

[⁸⁰ Pg 65] Bussard's idea for the ramjet appeared over 40 years ago [66]. Since then, various authors have made proposals and suggestions for the improvement of the initial ramjet design.

[⁸¹ Pg 66] The American physicist Robert Lull Forward (1932–), like many of the scientists mentioned in this book, is also a successful SF writer.

[⁸² Pg 66] See [67] for a discussion of laser sails in a one-way colonization mission; for there-and-back trips, see [68].

[⁸³ Pg 67] Stanislaw Marcin Ulam (1909–1984), a Polish-born mathematician, contributed to a number of fields. His autobiography [69] is fascinating. (Ulam appears in figure 28 on page 80.) The English-born physicist Freeman John Dyson (1923–) is one of the most imaginative physicists of his generation. For the papers on gravitational propulsion, see [70].

[⁸⁴ Pg 67] For a look at the possibilities of negative mass, see [71].

[⁸⁵ Pg 69] The American astronomer Carl Edward Sagan (1934–1996) based the science in his novel *Contact* on work by the American theoretician Kip Stephen Thorne (1940–) who has been prominent in investigating the properties of wormholes. In 1997, Sagan's novel was made into a movie of the same name, starring the excellent Jodie Foster.

[⁸⁶ Pg 70] Miguel Alcubierre Moya (1964–), a Mexican theoretical physicist, now works at the Max Planck Institute for Gravitational Physics in Potsdam, Germany. His paper describing the warp drive is in [72].

[[87] Pg 70]   For details on the possibility of using wormholes for transport, see [73]. For details on Van Den Broeck's warp drive, see [74]. These matters have been covered in detail, and at a non-mathematical level, in John Cramer's "Alternate View" columns in *Analog*; see http://www.npl. washington.edu/AV/ for past columns.

[[88] Pg 71]   In 1948, the Dutch physicist Hendrik Brugt Gerhard Casimir (1909–2000) predicted that quantum fluctuations of the EM field would cause a small attractive force to act between two close parallel uncharged conducting plates. The best direct measurement of this effect [75] used gold-coated quartz surfaces as the plates, with a torsion pendulum attached to one of the plates in such a way that if one plate moved toward the other, the pendulum would twist. A laser measured the twisting of the pendulum to an accuracy of 0.01 micron. The experiment confirmed Casimir's predictions. For articles propounding the idea that mankind might one day mine the zero-point energy see, for example, [76].

SOLUTION 10:   THEY HAVE NOT HAD TIME TO REACH US

[[89] Pg 72]   One of the first responses to Hart's paper was [77], in which it is argued a temporal explanation of the paradox is indeed valid.

[[90] Pg 73]   See [78]. In [79], Jones has written a particularly entertaining discussion of various colonization processes, from past human expansions through to possible human settlement of the Solar System and nearby stars.

[[91] Pg 73]   See [80].

[[92] Pg 73]   In physics, diffusion is a random molecular process, whereby energy or matter flows from a higher concentration to a lower concentration until a uniform distribution is attained. For example, if you heat one end of a rod, then the heat diffuses from the hot end to the cool end. The rate of the diffusion process depends upon the rod's material; in a metal rod, the diffusion is quick; in an asbestos rod, the diffusion is slow. Another example of a diffusion process occurs when you put a sugar lump in a cup of tea; unless you stir the tea, sugar molecules diffuse only slowly through the liquid. A solid can even diffuse into another solid: if gold is plated on copper, the gold diffuses into the surface of the copper — though it takes thousands of years for gold atoms to penetrate more than a tiny distance.

[[93] Pg 74]   See [81] by Ian Crawford for a well-written account of Galactic colonization models and their relation to the Fermi paradox. See [48] for details of Fogg's particular model of Galactic colonization.

SOLUTION 11:   A PERCOLATION THEORY APPROACH

[94 Pg 74]   Geoffrey Alan Landis (1955–), an American physicist who works at NASA, is another scientist who is better known as an SF writer. For details of his approach, see [82].

[95 Pg 75]   A probability p must, by definition, lie in the range between 0 and 1. A probability of $p = 0$ corresponds to an event that is impossible; a probability of $p = 1$ corresponds to an event that is certain to happen. If an event has only two outcomes — either the event happens or it does not — then the probability of the outcomes must add up to 1. (It is certain that *something* occurs!) So if the probability of the event happening is p, the probability of it not happening is $1 - p$.

[96 Pg 75]   Percolation theory was developed in 1957 by the British mathematician John Michael Hammersley (1920–) and his colleagues. See [83] for the best introduction to the ideas of percolation theory; however, although this excellent book is entertaining reading, readers should be aware that it inevitably contains an element of mathematics.

[97 Pg 76]   In general, the value of p_c cannot be derived analytically. Instead, we must use computer simulations to estimate p_c for a given system. An infinite square lattice, for example, has a value of p_c at about 0.592 75. A simple example should make clear the importance of a spanning cluster. Imagine a large chunk of some electrical insulating material, in which we embed a certain fraction, by volume, of identical electrically conducting spheres. Below the critical value p_c, no spanning cluster exists and the material remains an insulator. Above the critical value p_c, a spanning cluster exists and the material can conduct electricity. The same considerations tell us the density of people at which a disease will spread, or the density of trees at which a fire will consume an entire forest.

SOLUTION 12:   BRACEWELL–VON NEUMANN PROBES

[98 Pg 80]   It is appropriate to provide a website address for a reference on the history of computing! The National Archive for the History of Computing, a comprehensive British site hosted by Manchester University (the real, physical archive is also there) is at:

http://www.man.ac.uk/Science_Engineering/CHSTM/nahc.htm

[99 Pg 81]   The Australian-born electrical engineer Ronald Newbold Bracewell (1921–) has long been a leading light in SETI. See [84].

Chapter 7

Solution 13: We Are Solar Chauvinists

[100 Pg 84]   This resolution to the Fermi paradox was discussed in [85], a book that is now sadly out of print.

[101 Pg 84]   The concept of the Dyson sphere first appeared in [86]. (A Dyson sphere is a *loose* collection of bodies moving on independent orbits around a star; a rigid sphere would be unstable.) The idea inspired two great SF novels: *Ringworld* by Larry Niven and *Orbitsville* by Bob Shaw.

Solution 14: They Stay at Home...

[102 Pg 85]   The American astronauts Neil Alden Armstrong (1930–) and Edwin Eugene Aldrin Jr. (1930–) landed at the edge of Mare Tranquillitatis on 20 July 1969; Armstrong walked on the Moon at 10:56 P.M. (Eastern Daylight Time). The *last* man to walk on the Moon was Eugene Andrew Cernan (1934–), and unfortunately he seems set to hold this honor for a long while yet. He recounts his experiences of the Apollo program in [87].

[103 Pg 85]   The US businessman Denis Tito paid $20 million to the Russian space program for the privilege of becoming the first space tourist. It is a mystery to me why NASA has not embraced space tourism. Robert Heinlein imagined the possibilities a long time ago.

[104 Pg 86]   "Inconstant Moon," one of the finest stories by the American author Laurance (Larry) van Cott Niven (1938–), describes the events of a single night when the full moon shines brighter than ever before. It is a gem, and deservedly won the 1972 Hugo award for best short story.

[105 Pg 86]   See [88].

Solution 15: ...and Surf the Net

[106 Pg 88]   Set a billion years in the future, Arthur Clarke's novel *The City and the Stars* [89] imparts a sense of wonder and magnificent scope few novels can match. It also presents at least two explanations of the Fermi paradox, including the notion that mankind might prefer to stay in the "City" — safe from facing the realities of a harsh Universe.

Solution 16: They Are Signaling But We Do Not Know How to Listen

[107 Pg 90]   See [90]. A more recent search at 203 GHz of 17 stars known to produce excess infrared radiation (and thus perhaps host Dyson spheres) found nothing unusual; see [91].

[^{108} Pg 90]   At the famous Byurakan conference on communication with extraterrestrial intelligence, the American computer scientist Marvin Lee Minsky (1927–) pointed out that truly advanced energy-conscious ETCs might radiate at a temperature just above the cosmic background.

[^{109} Pg 90]   Whitmire and Wright [92] were not the first to suggest the stars themselves could be used to send signals. Philip Morrison (1915–) suggested the "eclipse" method 20 years earlier, and Drake had made similar suggestions before. But their paper is perhaps the first to give detailed calculations of how to modify stellar spectra to send a signal.

[^{110} Pg 91]   See [93], page 245.

[^{111} Pg 92]   Einstein's theory of general relativity predicted the existence of gravitational waves — ripples in spacetime. Such waves were demonstrated indirectly by the American physicists Joseph Hooten Taylor Jr. (1941–) and Russell Alan Hulse (1950–) through exquisitely accurate observations of PSR 1913+16. This pulsar is part of a binary system, its partner being another neutron star. As the two stars orbit each other, they lose energy in precisely the manner predicted by general relativity: the binary system is radiating gravitational energy in the form of waves. Astronomers hope the latest generation of detectors, such as LIGO (Laser Interferometer Gravitational-wave Observatory) will soon observe gravitational waves directly. Even LIGO, though, will be able to detect waves from only the most violent astronomical phenomena.

[^{112} Pg 92]   The American chemist Raymond Davis Jr. (1914–) has been running his solar neutrino experiment for more than 30 years. See [94].

SOLUTION 17:   THEY ARE SIGNALING BUT WE DO NOT KNOW AT WHICH FREQUENCY TO LISTEN

[^{113} Pg 96]   The Italian physicist Giuseppe Cocconi (1914–) worked at Cornell University with Morrison before returning to Europe to work at CERN, where he eventually became Director. His paper with Morrison [95] is one of the classics in SETI.

[^{114} Pg 96]   The hertz (Hz), a unit of frequency named after the German physicist Heinrich Rudolf Hertz (1857–1894), corresponds to one cycle of vibration per second. 1 MHz is 1 million vibrations per second; 1 GHz is 1 billion vibrations per second.

[^{115} Pg 98]   For suggestions of other SETI frequencies see [96], [97] and [98].

[^{116} Pg 99]   See [99] for a discussion of the Wow signal.

[^{117} Pg 100]   Paul Horowitz (1942–), a Harvard astronomer, has been at the forefront of SETI research for several years. Much of the funding for META came from Steven Spielberg (1947–), the director of ET.

[^{118} Pg 100]   The idea for SERENDIP came from the American astronomers Jill Tarter (1944–) and C. Stuart Bowyer (1934–) in 1978. Tarter, who is currently director of Project Phoenix and who holds a Chair at the SETI Institute, is widely believed to have been the inspiration for Sagan's heroine in *Contact*.

[^{119} Pg 100]   The American physicists Arthur Leonard Schawlow (1921–) and Charles Hard Townes (1915–) both won the Nobel Prize for physics (Townes in 1964 and Schawlow in 1981). Townes was far-seeing in regard to the potential of lasers, but few believed him. The suggestion that SETI should consider optical searches is almost as old as the Cocconi–Morrison paper: see [100].

[^{120} Pg 101]   That optical SETI is increasing in importance is largely due to the efforts of the British electrical engineer Stuart A. Kingsley (1948–). Kingsley has promoted the attractions of optical communication channels for more than a decade, and the astronomical community is finally coming round to his way of thinking.

[^{121} Pg 101]   In addition to looking for short laser pulses, as Kingsley does, astronomers have looked for other evidence for artifacts in the visible spectrum. One experiment [101] looked for the spectral lines of lasers, for example. Another looked for optical artifacts caused by astroengineering projects. Over the next few years we can expect optical SETI to become increasingly sophisticated.
    A great deal of information regarding all aspects of SETI can be found on the Web. For optical SETI, try [102]. The SETI Institute [103] has information about Project Phoenix. See [104] for information on Project BETA. To become involved with Project Argus, which aims to coordinate the efforts of amateur radio astronomers for the purposes of SETI, see [105].

[^{122} Pg 102]   See [106]. The author overstates his case, but the article is nevertheless accessible and thought-provoking.

SOLUTION 18:   OUR SEARCH STRATEGY IS WRONG

[^{123} Pg 102]   The SETI League development is an amateur project. Throughout history, the observations of amateurs have made important contributions to astronomy. It is extremely pleasing that amateur radio astronomers are now occupying a useful niche in the search for ETCs.

[^{124} Pg 103]   The analysis [107] was performed by telecommunications expert Nathan L. Cohen and computer scientist Robert Hohlfeld.

[^{125} Pg 104]   The "universal" frequency standard was first published in [108]. See also [109].

[^{126} Pg 104]   The SETI@home project [110] was founded by the American astronomer David Gedye (1960–). The idea behind it — namely, distributing small parts of a large computational problem to many processors — will be used increasingly in the future. Physicists are already working on a successor to the Internet, known as the Grid, which will be optimized for distributed processing. The possibilities are exciting.

[^{127} Pg 105]   Computing power can be rated in terms of FLOPS — floating point operations per second — a unit that measures the number of arithmetic processes that a computer can perform in 1 second. At the time of writing, the world's most powerful supercomputer is IBM's ASCI White, which has a rating of 12 TeraFLOPS: it can perform 12 trillion arithmetic operations per second. The SETI@home project is currently rated at 15 TeraFLOPS, and yet it cost a fraction of the price of the IBM machine. In September 2001, the project completed a world-record 10^{21} floating point operations — a ZettaFLOP!

SOLUTION 19:   THE SIGNAL IS ALREADY THERE IN THE DATA

[^{128} Pg 105]   From about 60 trillion signals, META researchers have found only 11 good candidate signals. If these signals were really attempts at communication, however, why could they not be re-observed? One suggestion was that interstellar plasmas or gravitational microlenses, passing between the sources and Earth, caused the steady signals to "twinkle" — and temporarily become strong enough for us to detect. Unfortunately, a recent analysis of the data has ruled out this possibility. This new result seems to indicate that the Galaxy contains, at most, just one other civilization with a comparable level of technology to ours that is deliberately trying to contact us. See [111]

[^{129} Pg 105]   The American astronomers Benjamin Michael Zuckerman (1943–) and Patrick Edward Palmer (1940–) surveyed 600 of the nearest Sun-like stars at 1420 MHz, but found no signals.

SOLUTION 20:   WE HAVE NOT LISTENED LONG ENOUGH

[^{130} Pg 106]   Drake wrote this in the Preface to *Is Anyone Out There?* [12].

[^{131} Pg 107] Of more than 100,000 respondents to a SETI@home poll, 89% believe the discovery will happen within the next 100 years. Almost half believe the discovery will happen within the next ten years. See the SETI@home web site for up-to-date details of the poll responses; the values I give here refer to November 2001.

[^{132} Pg 107] It is easy to produce large estimates for the number of communicating civilizations in the Galaxy: simply put "optimistic" values for the various factors into the Drake equation and you can produce numbers for N that are as large as 10^6.

SOLUTION 21:   EVERYONE IS LISTENING, NO ONE IS TRANSMITTING

[^{133} Pg 108] If ETCs could detect our television transmissions, then they could deduce a great deal about our planet even without decoding the programs. In 1978, the American astronomer Woodruff T. Sullivan III (1944–) showed how an ETC could deduce the rotational speed of Earth, estimate its size, the length of our year, the distance of Earth from the Sun, and the Earth's surface temperature!

[^{134} Pg 110] For more information on *Hipparcos* see [59].

[^{135} Pg 110] The idea that we could signal ETCs is almost 200 years old. In 1820 the German mathematician Johann Karl Friedrich Gauss (1777–1855), one of the greatest of all mathematicians, suggested planting forests of pine trees in such a way that they illustrated the Pythagorean theorem; this would signal our presence to any intelligent beings in the Solar System. The idea was expanded upon by Joseph Johann von Littrow (1781–1840), director of the Vienna Observatory, who suggested digging large ditches with geometrical shapes, filling them with kerosene, and setting them ablaze. He believed that light from these plainly artificial fires would be visible throughout the Solar System. In 1869, the French physicist Charles Cros (1842–1888) suggested that reflecting sunlight toward Mars using suitably arranged mirrors would be the best way to signal our presence to Martian astronomers.

[^{136} Pg 111] Yvan Dutil and Stephane Dumas, who work at the Canadian Defense Research Establishment, encoded a message in LINCOS and used the Evpatoria transmitter in Ukraine to send their message. The message was a series of "pages" describing some basic mathematics, physics and astronomy. The Dutil–Dumas experiment was promoted by an organization called *Encounter 2001*. You can find out more about the "cosmic call" experiment, and *Encounter 2001*, from the website in [112].

[^{137} Pg 111] For a discussion of this suggestion, as well as for general SETI questions, see [113].

SOLUTION 22: BERSERKERS

[^{138} Pg 112] The American author Fred Thomas Saberhagen (1930–) has written many stories about berserkers, the first collection appearing in *Berserker* in 1967. The concept of a Doomsday weapon was brilliantly satirized by Stanley Kubrick in *Dr. Strangelove*, and the original *Star Trek* television series aired an episode called *The Doomsday Machine*, which dramatized the notion of an indestructible world-killing machine (though Kirk and Co. managed to destroy it, of course). The machine in *Star Trek* was a single, large, slow-moving object. My mental picture of berserkers is somewhat different: I imagine swarms of small, fast-moving machines. A novel entitled *The Unreasoning Mask*, by the American author Philip José Farmer (1918–), is another that treats the notion of world-killers. But perhaps the idea of malignant killing machines has been treated most thoroughly by the American astrophysicist Gregory Benford (1941–), who is also one of the finest modern SF writers.

SOLUTION 23: THEY HAVE NO DESIRE TO COMMUNICATE

[^{139} Pg 113] Drake tells the story of how the English astronomer Martin Ryle (1918–1984), an Astronomer Royal and winner of the Nobel Prize for physics, was distraught upon learning of the 1974 Arecibo transmission toward M13. Ryle was worried that advanced ETCs might prey upon us.

My favorite fictional description of a species whose defining trait is extreme caution — taken to the point of cowardice — is that of "Puppeteers." They occur in Larry Niven's "Known Space" stories, including *Ringworld*.

[^{140} Pg 114] The two emperors mentioned here were Hongwu (1328–1398) and Yongle (1359–1424); the incredible voyages of the admiral Zheng He (c. 1371–c. 1435) have only relatively recently come to light.

[^{141} Pg 115] See [98].

[^{142} Pg 115] See [12, p. 210].

SOLUTION 24: THEY DEVELOP A DIFFERENT MATHEMATICS

[^{143} Pg 116] See [114].

[^{144} Pg 116]   For a wonderful critique of what animals may be doing when we say they are counting, see [115]. The book gives a superb introductory account of animal cognitive processes. For a critique of the Platonic view of mathematics, see [116]. A strong anti-Platonic case is presented in [117].

[^{145} Pg 117]   For a powerful argument as to why we *should* be able to converse with aliens, using our system of mathematics and perhaps a language like LINCOS, see [118].

[^{146} Pg 117]   The Argentinian writer Jorge Luis Borges (1899–1986), perhaps the greatest Spanish-language writer of the last century, was one author who could imagine alien mathematics, and his stories are a delight.

SOLUTION 25:   THEY ARE CALLING BUT WE DO NOT
RECOGNIZE THE SIGNAL

[^{147} Pg 119]   The LINCOS language was developed by the German mathematician Hans Freudenthal (1905–1990). There are a few websites devoted to LINCOS, but if you really want to learn the language I believe there is only one source: the original (but out of print) book [119].

[^{148} Pg 119]   If EM radiation is used to transmit information, the most efficient format for a given message is indistinguishable from black-body radiation (to a receiver who is unfamiliar with the format). This was first shown in [120]. The same result, using different arguments, was derived in [121].

[^{149} Pg 119]   The best print resource for the mysterious Voynich Manuscript is a small-press book [122], which is difficult to find. However, several websites describe the puzzle.

SOLUTION 26:   THEY ARE SOMEWHERE BUT THE UNIVERSE IS
STRANGER THAN WE IMAGINE

[^{150} Pg 121]   The American writer Alfred Bester (1913-1987) published his famous novel *The Stars My Destination* in 1956 [123]. Arthur Clarke's most ambitious work is perhaps *Childhood's End* [124]. Seemingly *outré* speculations are not limited to science fiction, however. Theoretical physicists also delight in dreaming up wild ideas; see, for example, [125].

[^{151} Pg 121]   Hugh Everett III (1930–1982) developed the many-worlds interpretation of quantum mechanics for his PhD thesis at Princeton. A summary of the thesis was published in [126]. Unfortunately his ideas were not taken seriously at the time of publication, and he became dispirited and left academia.

SOLUTION 27: A CHOICE OF CATASTROPHES

[152 Pg 122] Suggestions like this one, which are based on projecting human motivations and modes of thought onto alien minds, seem to me to show a lack of imagination. If we ever meet an alien intelligence, I believe it will *truly* be alien, with motivations that we will find hard to decipher.

[153 Pg 123] Drake and Sobel [12, p. 211] report how Shklovsky lost heart in the SETI enterprise in the years before his death. Shklovksy was convinced that nuclear war was inescapable, and the same inevitable holocaust would occur with other technological civilizations.

[154 Pg 123] One hesitates to use the word "intelligent" in this context, but the meaning is clear.

[155 Pg 123] See [127].

[156 Pg 123] Walter Michael Miller Jr. (1923–1996) was an American radioman and tailgunner on 53 bombing raids over Italy and the Balkans in World War II. His award-winning *A Canticle for Liebowitz* [128] is a classic post-apocalyptic SF novel. He wrote it in response to the Allied attack on Monte Cassino — a raid in which he took part and which almost certainly affected him psychologically. (Miller's post-holocaust world is vividly described, but do not read the book for scientific accuracy. Furthermore, the details of nuclear winter were only determined quite recently.)

[157 Pg 126] The term "nanotechnology" was popularized by the American physicist K. Eric Drexler. In an influential book [129] he presented his vision of a forthcoming revolution in nanoscale engineering. Drexler introduced the term "nanotechnology" to refer to molecular manufacturing (the construction of objects to complex, atomic specifications using sequences of chemical reactions directed by non-biological molecular machinery) together with its techniques, its products, and their design and analysis. Recently, the word "nanotechnology" has come to denote any technology that has nanoscale effects — submicron lithography (or etching) for example. To distinguish his original concept from the work currently taking place in laboratories, Drexler now refers to "molecular nanotechnology."

[158 Pg 127] See [130] for a detailed mathematical assessment of the environmental risks of nanotechnology.

[159 Pg 128] A website run by the physics department at the University of California at Davis [131] contains links to the original articles that sparked the flurry of controversy over the operation of RHIC, along with links to articles that quantify the risk and show it to be essentially zero.

Chapter 7

[^{160} Pg 129]   The smallest possible black hole is about 10^{-35} m across (the so-called Planck length); smaller structures are wiped out by quantum fluctuations). The creation of even the smallest black hole would require energies of around 10^{19} GeV, which is billions of times larger than the energies that our particle accelerators can generate. And even if we *could* create such a black hole, it would evaporate in 10^{-42} seconds. There are certainly more pressing things to worry about.

[^{161} Pg 129]   The existence of strange quarks has been known for decades. Their key properties were first highlighted by George Zweig (1937–) and Murray Gell-Mann (1929–) in 1964, but their presence was first evident in cosmic-ray experiments performed by Clifford Charles Butler (1922–1999) and George Rochester (1909–2001) as far back as 1947. It is an injustice that they did not receive a Nobel Prize for this achievement.

[^{162} Pg 129]   These calculations were the work of the American physicist Robert Loren Jaffe (1946–) and others. For a non-technical account, see [132]. For a more in-depth analysis, see [133].

[^{163} Pg 129]   The idea [134] that our Universe may not be in the "true" vacuum did not originate from cranks! Martin John Rees (1942–), an English astrophysicist, is the Astronomer Royal. His Dutch colleague Piet Hut (1952–) works at the Princeton Institute for Advanced Studies.

[^{164} Pg 130]   Fermilab's management became so exasperated with Dixon's protests that they discussed the matter in the 19 June 1998 issue of the newsletter *FermiNews* [135].

Kurt Vonnegut, in his novel *Cat's Cradle*, gives a fictional account of the effects of a phase transition (albeit a phase transition involving imaginary "ice-nine" — a form of H_2O that is more stable than ordinary water at room temperature — rather than the vacuum).

[^{165} Pg 130]   On 15 October 1991 the Fly's Eye detector in Utah detected a cosmic ray with an energy of 320 EeV. (This energy is so large that the rarely-used SI prefix "Exo" was required; the prefix represents a factor of 10^{18}.) The particle detected by Fly's Eye packed a *staggering* amount of energy: about 50 J. In other words, this single subatomic particle carried more kinetic energy than a tennis ball traveling at 180 mph. Its energy was more than 10 million times greater than the maximum achievable energy of the largest accelerator ever been planned. How this particle acquired so much energy is something of a mystery. No obvious process can produce a particle with this much kinetic energy; yet *whatever* produced it must have been relatively nearby, because if it had traveled cosmological distances its interactions with the microwave background would have slowed it down. See [136].

[[166] Pg 130]   J. Richard Gott III (1947–) is a professor of astrophysics at Princeton University. His original paper on the Doomsday argument [137] purported to show, among other things, that mankind is unlikely to colonize the Galaxy. The article generated an interesting correspondence [138]. The philosopher John Leslie independently developed the Doomsday argument [139]. Perhaps the first person to appreciate the power of this type of reasoning was the English physicist Brandon Carter (1942–); Carter's anthropic arguments are outlined in Chapter 5.

SOLUTION 28:   THEY HIT THE SINGULARITY

[[167] Pg 134]   Gordon E. Moore (1929–) co-founded Intel in 1968.

[[168] Pg 135]   Moore's law, rather than miserliness, is the main reason why I am loath to upgrade my computer. I figure that if I wait for six more months I will get something much better for my money; on the other hand, it means I have been waiting for five years now to upgrade.

[[169] Pg 135]   The American mathematician Vernor Steffen Vinge (1944–) has explored the idea of the Singularity in several SF novels and short stories. A non-fictional account of the idea can be found in [140].
    A superb discussion of the seemingly inexorable development of computing power can be found in [141].

[[170] Pg 135]   The term "singularity" was used in the 1950s by von Neumann. He said: "The ever accelerating progress of technology ... gives the appearance of approaching some essential singularity in the history of the race beyond which human affairs, as we know them, could not continue."

[[171] Pg 135]   Vinge was not the first to explore the idea that mankind's intellectual development might profoundly change our global society. The French Jesuit priest Pierre Teilhard de Chardin (1881–1955) thought individual minds would somehow merge to form the noösphere — an expanding sphere of human knowledge and wisdom; spiritual and material would eventually merge to form a new state of consciousness he called the Omega point. His argument, although mystical and woolly, reaches a conclusion that seems similar to Vinge's Singularity. There are two main differences between Vinge and Teilhard de Chardin. First, Vinge has extrapolated real-world trends to suggest specific mechanisms that might get us to the Singularity. Second, organic evolution requires millions of years to construct the noösphere; we (and our successors) construct the Singularity in a few decades.

Chapter 7

[^{172} Pg 136]   See [142] for two stimulating books criticizing the idea that human-level "artificial" intelligence can exist. Personally, I disagree with the conclusions of these distinguished thinkers; but the two references here make for extremely interesting reading.

[^{173} Pg 136]   TeX was developed by the American computer scientist Donald Ervin Knuth (1938–). He wrote the software (along with a program for designing typefaces) just so that he could typeset his multi-volume *Art of Computer Programming* to his own satisfaction! See [143].

SOLUTION 29:   CLOUDY SKIES ARE COMMON

[^{174} Pg 138]   Asimov's story "Nightfall" is routinely voted as the best SF short story of all time. It can be found in many collections, including [144].

SOLUTION 30:   INFINITELY MANY ETCS EXIST BUT ONLY ONE WITHIN OUR PARTICLE HORIZON: US

[^{175} Pg 138]   Hart is a particularly clear and forceful writer. For a description of his proposal of how an infinite number of life-bearing planets exist, yet we are alone in the observable Universe, see [145]. An equally clear treatment of the subject, by a cosmologist, appears in [146].

CHAPTER 5   THEY DO NOT EXIST

[^{176} Pg 141]   The thought-provoking book by Ward and Brownlee [147] articulates the growing suspicion of a number of astrobiologists that Earth is unusual, perhaps unique, in harboring complex life-forms.

[^{177} Pg 142]   For an imaginative, unorthodox and challenging book on the possible forms that life may take, see [148]. The authors discuss the notions of plasma life in stars, radiant life in interstellar clouds, silicate life, low-temperature life and many other possibilities. One of the earliest and most delightful SF stories about alien biochemistries was *A Martian Odyssey* by Stanley G. Weinbaum (in *Wonder Stories*, July 1934). You can find the story in several anthologies, including [149].

SOLUTION 31:   THE UNIVERSE IS HERE FOR US

[^{178} Pg 143]   See, for example, [150].

[^{179} Pg 146]   See [151].

[180 Pg 146]   See [152] — a remarkable and stimulating book.

[181 Pg 146]   See [25].

SOLUTION 32:   LIFE CAN HAVE EMERGED ONLY RECENTLY

[182 Pg 147]   See [153].

SOLUTION 33:   PLANETARY SYSTEMS ARE RARE

[183 Pg 150]   The novels mentioned in the text were *Integral Trees* by Larry Niven and *Dragon's Egg* by Robert Forward.

[184 Pg 150]   The French naturalist George-Louis Le Clerc, Comte de Buffon (1707–1788), proposed in 1749 that the planets formed when a comet collided with the Sun. The German philosopher Immanuel Kant (1724–1804) proposed the nebular theory of planetary formation in 1754.

[185 Pg 151]   The earliest models of planetary formation by stellar collisions were developed by the American scientists Thomas Chrowder Chamberlin (1843–1928) and Forest Ray Moulton (1872–1952). The models were changed and improved by the British mathematicians James Hopwood Jeans (1887–1946) and Harold Jeffreys (1891–1989). See [154] for a fascinating tour of the Solar System, including its formation. The author reaches the conclusion that life on Earth may be the result of chance; and perhaps this means that life is unlikely to occur elsewhere.

[186 Pg 152]   For more details on the newest planetary discoveries, visit *The Extrasolar Planets Encyclopædia* — a website run by Jean Schneider [155].

SOLUTION 34:   WE ARE THE FIRST

[187 Pg 154]   News of the new planet orbiting 47 Ursae Majoris was announced on 15 August 2001. For details, see [156]. You can also find out more about the planet in [155].

SOLUTION 35:   ROCKY PLANETS ARE RARE

[188 Pg 156]   The accepted age of the Earth, as calculated by geochemists using radioisotopic dating techniques, is 4.55 ± 0.7 billion years. This value was first presented in 1956 by the American geochemist Clair Cameron Patterson (1922–1995). Thus, within the margins of error, we can be confident that Earth formed at the same time as the chondrites.

[189 Pg 157]   References to what we now know are chondrules were made
in the scientific literature as far back as 1802, though they were not named
until 1864 (by the German mineralogist Gustav Rose [1798–1873]). The English geologist Henry Clifton Sorby (1826-1908), one of the great amateur
scientists, used a petrographic microscope — a device he invented — to
carry out the first detailed study of chondrules. He suggested that chondrules, which he described as being "like drops of a fiery rain," might be
pieces of the Sun that had been ejected in solar prominences.

[190 Pg 157]   See [157].

SOLUTION 36:   CONTINUOUSLY HABITABLE ZONES ARE NARROW

[191 Pg 158]   One of the first books to discuss the conditions that might be
required to make a planet habitable for mankind was [158]. Although now
rather dated, it remains an excellent guide to the problems that must be
faced. The book was the outcome of a RAND study and is rather technical.
A popular version, also recommended, is [159].

[192 Pg 158]   In most models, an Earth-like planet freezes when a CO_2
"blanket" prevents radiation from its star penetrating the atmosphere.

[193 Pg 159]   See [160].

[194 Pg 159]   The American geologist James Fraser Kasting (1953–) has
made several important contributions to our understanding of the long-
term stability of the Earth's climate. The models he and his co-workers
employ are much more detailed than Hart's original model. See [161] for
further details.

[195 Pg 160]   See [162].

SOLUTION 37:   JUPITERS ARE RARE

[196 Pg 163]   When Earth first condensed from the protoplanetary disk,
temperatures were too high for it to have retained water. So our oceans of
water must have been delivered *after* Earth had cooled. If the water came
when Earth was already at, or close to, its present mass, then it would be
massive enough to keep hold of most of its water. But where did all the
water come from? The standard scenario is that the water condensed into
ice in the outer regions of the disk — perhaps in comets, where temperatures were cooler. A cometary bombardment later delivered the oceans.
Recent work casts this scenario into doubt. We know from measurements
of Jupiter that the initial solar nebula contained about 30 parts per million of deuterium, and measurements of comets Hale–Bopp, Halley and

Hyakutake demonstrate that comets contain about 450 parts per million of deuterium. Neither of these values are close to the value for sea water, which contains about 150 parts per million of deuterium. Meteorites from the outer Asteroid Belt, however, have the same deuterium abundance as does sea water. It thus seems probable that Earth obtained its water from a collision with a large planetary embryo, rather than from cometary bombardment. See [163] for further details.

SOLUTION 38:   EARTH HAS AN OPTIMAL "PUMP OF EVOLUTION"

[197 Pg 164]   See [164].

[198 Pg 164]   That resonance effects should cause gaps to exist in the Asteroid Belt was first suggested in 1866 by the American astronomer Daniel Kirkwood (1814–1895). Jack Leach Wisdom (1953–), an American physicist, was one of the first scientists to apply the modern techniques of nonlinear dynamics to the study orbits in the Solar System. Wisdom looked at the Asteroid Belt's 3:1 resonance in detail. The American geologist George West Wetherill (1925–) is well known for his research into the role that Jupiter plays in the Solar System.

SOLUTION 39:   THE GALAXY IS A DANGEROUS PLACE

[199 Pg 166]   Magnetars are neutron stars with exceptionally strong magnetic fields. The field of SGR 1900+14 is estimated to be 5×10^{14} Gauss; compare that with the strongest sustained magnetic field scientists have made, which is only 4×10^5 Gauss. The magnetic field of a magnetar is so strong that it could suck the keys from your pocket at a distance of more than 100,000 miles. Of course, if you were standing that close to a magnetar, then the radiation and charged-particle wind that it spews out would kill you instantly .

[200 Pg 170]   Gamma-ray bursters were first detected in 1969 by the VELA satellites (which were in orbit to look for gamma-rays from possible nuclear explosions), but it was not until 1997 that astronomers obtained proof that bursters occur at cosmological distances; even now the precise nature of the progenitor events is a matter for debate.

[201 Pg 172]   See [165].

[202 Pg 172]   Arthur Clarke's haunting short story "The Star" appears in many anthologies. See, for example, [166].

Chapter 7

SOLUTION 40:   A PLANETARY SYSTEM IS A DANGEROUS PLACE

[203 Pg 174]   The notion that Earth experienced a global glaciation in the Neoproterozoic age is not new: the English geologist Brian Harland postulated precisely this as long ago as 1964. At the same time, the Russian geologist Mikhail Budyko showed how a runaway icehouse effect could take place. Only recently, however, has the notion been taken seriously — largely due to the work of groups led by the American geologists Joseph Kirschvink and James Kasting, who have investigated the escape route from "Snowball Earth." For an early introduction, see [167]. A clearly written introduction to Snowball Earth theories appears in [168]. More technical papers include [169] and [170].

[204 Pg 176]   See [171].

[205 Pg 176]   See [172].

[206 Pg 177]   The idea that meteorite impact killed the dinosaurs is an old one. The key paper is [173]. Years before that paper appeared, however, a remarkably prescient article was published in an SF magazine [174]. It described the consequences of a large meteor hitting Earth. An entertaining look at the evidence for a meteorite impact causing the Cretaceous–Tertiary extinction appears in [175]; the book is as good as its title!

[207 Pg 180]   See [176].

SOLUTION 41:   EARTH'S SYSTEM OF PLATE TECTONICS IS UNIQUE

[208 Pg 180]   The earthquake of 17 August 1999 around Izmit, Turkey, resulted in many thousands of deaths and the destruction of countless homes. The death toll from the earthquake of 26 January 2001, which struck the Kachchh area of Gujarat, India, was more than 20,000. The Izmit earthquake was due to the catastrophic release of stress in the North Anatolian fault zone, where the Anatolian and Eurasian plates meet. The Indian earthquake — the most damaging to strike that region in more than 50 years — was caused by the Indian plate pushing northward into the Eurasian plate.

[209 Pg 181]   The first to marshal evidence for the suggestion that continents move was the German meteorologist Alfred Lothar Wegener (1880–1930). He published his ideas on continental drift in 1915, but they were met with ridicule. One of the seeming flaws in his theory was that no known mechanism could account for the drift of continents. Wegener died in a blizzard on an arctic expedition, shortly before the British geologist Arthur Holmes (1890–1965) suggested that convection might provide a suitable mechanism to explain continental drift. Holmes was a respected geologist; he was the first, for example, to suggest a reasonable timescale for geological processes. His 1913 estimate of 4 billion years for the age of the Earth was far better than any previous estimate. But it was to be almost another 20 years before the idea became established. In 1960, the American geologist Harry Hammond Hess (1906–1969) showed that the seafloor was spreading from vents in mid-ocean rifts. As magma welled up and cooled, it pushed the existing seafloor away from both sides of the rifts. It was this force that moved the continents.

[210 Pg 182]   The first description of Earth's geological-timescale CO_2 thermostat appeared in [177]. This mechanism does not take into account the effect that biological organisms might have had on stabilizing global surface temperature. Several prominent scientists take the view that life itself has played the key role in keeping temperature at an equable level.

SOLUTION 42:   THE MOON IS UNIQUE

[211 Pg 185]   Two groups independently arrived at the idea of lunar formation by a Mars-sized impactor. One group was led by the American astronomers William K. Hartmann (1939–) and Donald Ray Davis (1939–), who work at the Planetary Science Institute in Arizona. The other group was led by the Canadian-American astronomer Alastair Graham Walter Cameron (1925–) of Harvard University.

[212 Pg 187]   For an entertaining treatment of the importance of the Moon, which is aimed at non-scientists, see [178].

SOLUTION 43:   LIFE'S GENESIS IS RARE

[213 Pg 190]   The classification of living organisms into the domains of archaea, bacteria and eukarya is relatively recent. The proposal came from the American biophysicist Carl R. Woese (1928–), who discovered microorganisms living in extreme environments (extremes of heat, salinity, acidity — places previously thought to be hostile to life). At first it was thought that these organisms were bacteria that had managed to adapt to extreme conditions; certainly, the cell nucleus of these organisms was not enclosed within a nuclear membrane, which made them look like bacteria. However, Woese and co-workers embarked on a study of the ribosomal RNA of these extremophiles. (In cells, ribosomal RNA is the site of protein synthesis — the place where amino acids are assembled into proteins. It is thus found in all living cells, and a study of the nucleotide sequence of rRNA provides an ideal "evolutionary chronometer.") They found that the rRNA of extremophiles differs quite radically from the rRNA of bacteria. These and other fundamental differences made it clear to Woese that life consists of *three* domains. The landmark paper is [179].

[214 Pg 193] See [180]. The author argues persuasively that as-yet unknown principles of self-organization may underpin phenomena as diverse as the beginnings of life to the workings of market economies.

[215 Pg 194] The story of nucleic acids goes back a long way. The German biochemist Albrecht Kossel (1853–1927) was the first to investigate the chemical structure of the nucleic acid molecule; it was he who isolated the nitrogen bases and named them adenine, guanine, cytosine and thymine. He won the 1910 Nobel Prize for his discovery. Forty years later, the role that DNA might play in heredity was one of the burning issues of biology. In 1953, Francis Crick and James Watson made one of the key breakthroughs in all of science when they proposed the double-helix model of the DNA molecule.

[216 Pg 200] For a good, colorful undergraduate-level textbook on genetics see [181]; the chapters on gene expression and the regulation of gene activity are particularly useful. Another heavy-duty textbook is [182].

[217 Pg 200] See for example [183]. A good collection of scholarly articles dealing with the possible importance of comets to life on Earth, including the idea that comets might have transported amino acids and other necessary materials to Earth, can be found in [184].

[218 Pg 200] The story of scientific research into the question of life's origin is long and fascinating. It began in 1924 with the Russian biologist Alexander Ivanovich Oparin (1894–1980), who suggested that small lumps of organic matter might have formed naturally and become the precursor of modern proteins. Along with the British biologist John Burdon Sanderson Haldane (1892–1964), he produced the evocative idea of the primordial soup, from which living material arose. It was not until 1953 that the American biologist Stanley Lloyd Miller (1930–), a graduate student working in the laboratory of the Nobel Prize-winning chemist Harold Clayton Urey (1893–1981), put these ideas to an experimental test. The results of Miller's experiments suggested that at least the basic building blocks of life could form naturally on a primordial Earth. Nevertheless, there are many steps leading from these building blocks to life itself, and the route remains shrouded in fog. This is a fascinating and active area of research.

[219 Pg 201] For an argument as to why the emergence of life might be a rare occurrence, see [185]. I believe the arguments in the paper are wrong, but as usual Hart states his case clearly and forcefully.

[^{220} Pg 202]   The first ribozymes — enzymes made of RNA — were discovered independently in 1983 by the American biochemist Thomas Robert Cech (1947–) and the Canadian biochemist Sidney Altman (1939–).

[^{221} Pg 205]   For an entertaining look at the logic behind SETI activities, see [186]. A work pitched at a similar level is [187].

[^{222} Pg 205]   Two excellent books on the problem of the origin of life, and the probability of life arising elsewhere, are [188] and [189]. Both books contain technical material, but both can be appreciated by the general reader. The book by de Duve, in particular, is exceptionally thorough — and reaches the conclusion that life must be common in the Universe.

[^{223} Pg 205]   Several articles deal with the possibility of life on the satellites of the giant planets. The original suggestion was made in [190]. For more recent articles, which contain many further references, see [191] and [192].

SOLUTION 44:   THE PROKARYOTE–EUKARYOTE TRANSITION IS RARE

[^{224} Pg 210]   See [193].

SOLUTION 45:   TOOLMAKING SPECIES ARE RARE

[^{225} Pg 214]   There is a wide literature on animal tool use, though there is no single definition of what constitutes tool use. Is a dog using a wall as a tool when it scratches its back? Depending upon one's definition, many animals have been observed to use tools. With regard to chimps, for example, see [194] and [195]. With regard to capuchin monkeys, see [196]. With regard to elephants, see [197]. Three good books on the subject (including the development of human tool use) are [198], [199] and [200].

SOLUTION 46:   TECHNOLOGICAL PROGRESS IS NOT INEVITABLE

[^{226} Pg 215]   See [201] and [202].

[^{227} Pg 216]   An introductory article describing how various hominid species must once have co-existed is given in [203]. For four excellent books on early-human tool use, see [204], [205], [206] and [207].

[^{228} Pg 216]   For a discussion of cave art see, for example, [208].

SOLUTION 47:   INTELLIGENCE AT THE HUMAN LEVEL IS RARE

[^{229} Pg 222]   For articles on the *eyeless* gene in fruit flies, see [209] and [210].

Chapter 7

Solution 48: Language Is Unique to Humans

[^{230} Pg 223]   See [115] for a superb account of research into animal cognition. This is a wonderful book. For a different view regarding the question of animal consciousness and intelligence, see [211].

[^{231} Pg 223]   See [212] for a discussion of the relevance of human linguistic abilities to the Fermi paradox.

[^{232} Pg 224]   The American linguist Avram Noam Chomsky (1928–), one of the world's most respected intellectuals, writes widely on political and social issues as well as on linguistics. His linguistic work is highly abstruse, but for an introduction to the revolution that he sparked in 1959 — and to the advances made by others in the intervening decades — look no further than Pinker's superbly readable book [213].

[^{233} Pg 226]   As this book was sent to press, news was published of the first identification of a gene responsible for a specific language disorder. The gene, called FOXP2, encodes an uncommon protein involved in the development of neural structures in the embryo. Disruption of the FOXP2 gene leads to abnormality in neural structures that are important for speech and language. This discovery bolsters the notion that our ability to talk has a strong genetic component. See [214].

Solution 49: Science Is Not Inevitable

[^{234} Pg 232]   There are several good accounts of the historical development of science. See, for example, [215].

Chapter 6   Conclusion

[^{235} Pg 233]   Brin's article [18] was referred to above, in the section entitled "The Fermi Paradox."

Solution 50: The Fermi Paradox Resolved ...

[^{236} Pg 235]   Eratosthenes of Cyrene (c. 276 BC–c. 196 BC) was a man of wide-ranging interests. Other Greeks nicknamed him Beta, the second letter of the Greek alphabet, because he was the second-best scholar in so many fields of knowledge. For information about prime numbers, consult any book on numbers. My own favorite is [216].

[^{237} Pg 240]   See [217].

[^{238} Pg 240]   See [218]. This translation from the French original is by A. Wainhouse.

8

References

[1]   Fermi L (1954) *Atoms in the Family* (Chicago: University of Chicago Press)
[2]   Segré E (1970) *Enrico Fermi: Physicist* (Chicago: University of Chicago Press)
[3]   Poundstone W (1988) *Labyrinths of Reason* (London: Penguin)
[4]   Rapoport A (1967) Escape from paradox. *Scientific American* **217** (1) 50–6
[5]   Williamson T (1994) *Vagueness* (London: Routledge)
[6]   Gardner M (1969) *The Unexpected Hanging and Other Mathematical Diversions* (New York: Simon and Schuster)
[7]   Mermin N D (1990) *Boojums all the Way Through* (Cambridge: CUP)
[8]   Gribbin J (1996) *Schrödinger's Kittens* (London: Phoenix Press)
[9]   Einstein A, Podolsky B and Rosen N (1935) Can a quantum-mechanical description of physical reality be considered complete? *Physical Review* **41** 777–80
[10]   Harrison E (1987) *Darkness at Night* (Cambridge, MA: Harvard University Press)
[11]   Jones E M (1985) Where is everybody? An account of Fermi's question. *Preprint LA-10311-MS*. (Los Alamos, NM: Los Alamos National Laboratory). A more easily located version of this reference appeared in the August 1985 issue of *Physics Today* pp 11–3.)
[12]   Drake F and Sobel D (1991) *Is Anyone Out There?* (London: Simon and Schuster)
[13]   Lytkin V, Finney B and Alepko L (1995) Tsiolkovsky, Russian cosmism and extraterrestrial intelligence. *Quarterly J. Royal Astronomical Soc.* **36** 369–76
[14]   Viewing D (1975) Directly interacting extra-terrestrial technological communities. *J. Brit. Interplanetary Soc.* **28** 735–44
[15]   Hart M H (1975) An explanation for the absence of extraterrestrials on Earth. *Quarterly J. Royal Astronomical Soc.* **16** 128–35
[16]   Douglas (1977) The absence of extraterrestrials on Earth. *Quarterly J. Royal Astronomical Soc.* **18** 157–8
[17]   Tipler F J (1980) Extraterrestrial intelligent beings do not exist. *Quarterly J. Royal Astronomical Soc.* **21** 267–81

[18]   Brin G D (1983) The "great silence": the controversy concerning extraterrestrial intelligent life. *Quarterly J. Royal Astronomical Soc.* **24** 283–309
[19]   Brin G D (1985) Just how dangerous *is* the Galaxy? *Analog* **105** (7) 80–95
[20]   Zuckerman B and Hart M H (eds) (1995) *Extraterrestrials — Where Are They?* (Cambridge: CUP)
[21]   Aczel A (1998) *Probability 1: Why There Must be Intelligent Life in the Universe* (New York: Harcourt Brace)
[22]   Smolin L (1997) *The Life of the Cosmos* (London: Weidenfeld and Nicolson)
[23]   Gould S J (1985) SETI and the wisdom of Casey Stengel. In *The Flamingo's Smile* (London: Penguin)
[24]   Reinganum M R (1986–7) Is time travel impossible? A financial proof. *J. of Portfolio Management* **13**(1) 10–2
[25]   Tipler F J (1994) *The Physics of Immortality: Modern Cosmology, God and the Resurrection of the Dead* (New York: Anchor)
[26]   McPhee J (1973) *The Curve of Binding Energy* (New York: Farrar, Straus and Giroux)
[27]   Lanoutte W *et al* (1994) *Genius in the Shadows* (Chicago: University of Chicago Press)
[28]   Hall M D and Connors W A (2000) *Captain Edward J. Ruppelt: Summer of the Saucers* (Albuquerque, NM: Rose Press)
[29]   Sheaffer R (1995) An examination of claims that extraterrestrial visitors to Earth are being observed. In Zuckerman B and Hart M H (eds) *Extraterrestrials: Where Are They?* (Cambridge: CUP)
[30]   Story R (1976) *The Space Gods Revealed: A Close Look at the Theories of Erich von Däniken* (New York: Barnes and Noble)
[31]   Freitas R A Jr and Valdes F (1980) A search for natural or artificial objects located at the Earth–Moon libration points *Icarus* **42** 442–7. Also see Freitas R A Jr (1983) If they are here, where are they? Observational and search considerations. *Icarus* **55** 337–43
[32]   Lawton A T and Newton S J (1974) Long delayed echoes: the search for a solution. *Spaceflight* **6** 181–7
[33]   Lunan D (1974) *Man and the Stars* (London: Souvenir Press)
[34]   Sheehan W (1996) *The Planet Mars: A History of Observation and Discovery* (Tucson, AZ: University of Arizona Press)
[35]   Hoagland R C (1987) *The Monuments of Mars* (Berkeley, CA: North Atlantic Books)
[36]   Hancock G, Bauval R and Grigsby J (1998) *The Mars Mystery* (London: Michael Joseph)
[37]   Gardner M (1985) The great stone face and other nonmysteries. *Skeptical Inquirer* **9** (4)
[38]   Papagiannis M D (1978) Are we all alone, or could they be in the asteroid belt? *Quarterly J. Royal Astronomical Soc.* **19** 236–51
[39]   Stephenson D G (1978) Extraterrestrial cultures within the Solar System? *Quarterly J. Royal Astronomical Soc.* **19** 277–81
[40]   Freitas R A Jr (1983) The search for extraterrestrial artifacts (SETA) *J. Brit. Interplanetary Soc.* **36** 501–6. See also Freitas R A Jr (1985) There is no Fermi paradox. *Icarus* **62** 518–20
[41]   Yokoo H and Oshima T (1979) Is bacteriophage phi X174 DNA a message from an extraterrestrial intelligence? *Icarus* **38** 148–53
[42]   Nakamura H (1986) SV40 DNA: a message from Epsilon Eridani? *Acta Astronautica* **13** 573–8
[43]   Arrhenius S A (1908) *Worlds in the Making* (New York: Harper and Row)
[44]   Crick F H C and Orgel L E (1973) Directed panspermia. *Icarus* **19** 341–6. A fuller account is given in Crick F H C (1981) *Life Itself* (New York: Simon and Schuster)
[45]   Ball J A (1973) The zoo hypothesis. *Icarus* **19** 347–9

References

[46] Deardorf J W (1986) Possible extraterrestrial strategy for Earth. *Quarterly J. Royal Astronomical Soc.* **27** 94–101; Deardorf J W (1987) Examination of the embargo hypothesis as an explanation for the Great Silence. *J. Brit. Interplanetary Soc.* **40** 373–9

[47] Carey S S (1997) *A Beginner's Guide to Scientific Method* (Stamford, CT: Wadsworth)

[48] Fogg M J (1987) Temporal aspects of the interaction among the first galactic civilizations: the "interdict hypothesis." *Icarus* **69** 370–84

[49] Fogg M J (1988) Extraterrestrial intelligence and the interdict hypothesis. *Analog* **108** (10) 62–72

[50] Asimov I (1981) *Extraterrestrial Civilizations* (London: Pan)

[51] Newman W I and Sagan C (1981) Galactic civilizations: population dynamics and interstellar diffusion. *Icarus* **46** 293–327

[52] Baxter S (2000) The planetarium hypothesis: a resolution of the Fermi paradox. *J. Brit. Interplanetary Soc.* **54** 210–6

[53] Deutsch D (1998) *The Fabric of Reality* (London: Penguin). See also Tipler F J (1994) *The Physics of Immortality* (New York: Anchor)

[54] Smolin L (1997) *The Life of the Cosmos* (London: Weidenfeld and Nicolson)

[55] Harrison E (1995) The natural selection of universes. *Quarterly J. Royal Astronomical Soc.* **36** 193–203

[56] The *Voyager Project Page* is at http://voyager.jpl.nasa.gov

[57] The JPL site is at http://www.jpl.nasa.gov/index.cfm

[59] Webb S (1999) *Measuring the Universe* (London: Springer UK)

[58] French A P (1968) *Special Relativity* (San Francisco: Norton)

[60] Bernal J D (1969) *The World, The Flesh and the Devil* (Indiana: Indiana University Press). This is reprint of an earlier edition.

[61] Bova B (ed) (1973) *The Science Fiction Hall of Fame vol 2A* (New York: Doubleday)

[62] Anderson P (2000) *Tau Zero* (London: Orion). This is a re-issue of a classic book in the SF Collector's Edition.

[63] Hemry J G (2000) Interstellar navigation or getting where you want to go and back again (in one piece). *Analog* **121** (11) 30–7

[64] Mallove E F and Matloff G L (1989) *The Starflight Handbook* (Wiley: New York)

[65] Crawford I A (1995) Interstellar travel: a review. In Zuckerman B and Hart M H (eds) *Extraterrestrials: Where Are They?* (Cambridge: CUP)

[66] Bussard R W (1960) Galactic matter and interstellar flight. *Astronautica Acta* **6** 179–94

[67] Dyson F J (1982) Interstellar propulsion systems. In Zuckerman B and Hart M H (eds) *Extraterrestrials: Where Are They?* (Cambridge: CUP)

[68] Forward R L (1984) Roundtrip interstellar travel using laser-pushed lightsails. *J. of Spacecraft and Rockets* **21** 187–95

[69] Ulam S M (1976) *Adventures of a Mathematician* (Berkeley, CA: University of California Press)

[70] Ulam S M (1958) On the possibility of extracting energy from gravitational systems by navigating space vehicles. Report LA-2219-MS. (Los Alamos, NM: Los Alamos National Laboratory). See also Dyson F J (1963) Gravitational machines. In *Interstellar Communication* ed. A G W Cameron (New York: Benjamin)

[71] Forward R L (1990) The negative matter space drive. *Analog* **110** (9) 59–71

[72] Alcubierre M (1994) The warp drive: hyper-fast travel within general relativity. *Classical and Quantum Gravity* **11** L73–7

[73] Krasnikov S V (2000) A traversible wormhole. *Phys. Rev.* D **62** 084028

[74] Van Den Broeck C (1999) A "warp drive" with more reasonable total energy requirements. *Classical and Quantum Gravity* **16** 3973–9

[75] Lamoreaux S K (1997) Demonstration of the Casimir force in the 0.6 to 6 μm range. *Physical Review Letters* **78** 5–8

Chapter 8

[76]   Haisch B, Rueda A and Puthoff H E (1994) Beyond $E = mc^2$. *The Sciences* **34** (6) 26–31. See also Puthoff H E (1996) SETI, the velocity of light limitation, and the Alcubierre warp drive: an integrating overview. *Physics Essays* **9** 156

[77]   Cox L J (1976) An explanation for the absence of extraterrestrials on Earth. *Quarterly J. Royal Astronomical Soc.* **17** 201–8

[78]   Jones E M (1975) Colonization of the Galaxy. *Icarus* **28** 421–2; and Jones E M (1981) Discrete calculations of interstellar migration and settlement. *Icarus* **46** 328–36

[79]   Jones E M (1995) Estimates of expansion timescales. In Zuckerman B and Hart M H (eds) *Extraterrestrials: Where Are They?* (Cambridge: CUP)

[80]   Newman W I and Sagan C (1981) Galactic civilizations: population dynamics and interstellar diffusion. *Icarus* **46** 293–327

[81]   Crawford I A (2000) Where are they? *Scientific American* **283** (7) 28–33

[82]   Landis G A (1998) The Fermi paradox: an approach based on percolation theory *J. Brit. Interplanetary Soc.* **51** 163–6

[83]   Stauffer D (1985) *Introduction to Percolation Theory* (London: Taylor and Francis)

[84]   Bracewell R N (1960) Communication from superior galactic communities. *Nature* **186** 670–1

[85]   Rood R T and Trefil J S (1981) *Are We Alone?* (New York: Charles Scribner's)

[87]   Cernan E and Davis D (1999) *The Last Man on the Moon* (New York: St Martin's Press)

[88]   Zuckerman B (1985) Stellar evolution: motivation for mass interstellar migration. *Quarterly J. Royal Astronomical Soc.* **26** 56–9

[89]   Clarke A C (1956) *The City and the Stars* (New York: New American Library)

[86]   Dyson F J (1960) Search for artificial sources of infrared radiation. *Science* **131** 1667

[90]   Jugaku J and Nishimura S E (1991) A search for Dyson spheres around late-type stars in the IRAS catalog. In Heidemann J and Klein M J (eds) *Bioastronomy: The Search for Extraterrestrial Life (Lectures Notes in Physics)* **390** (Berlin: Springer)

[91]   Mauersberger R *et al.* (1996) SETI at the spin-flip line frequency of positronium. *Astronomy and Astrophysics* **306** 141–4

[92]   Whitmire D P and Wright D P (1980) Nuclear waste spectrum as evidence of technological extraterrestrial civilizations. *Icarus* **42** 149–56

[93]   Sullivan W (1964) *We Are Not Alone* (London: Pelican)

[94]   Bahcall J N and Davis R (2000) The evolution of neutrino astronomy. *CERN Courier* **40** (6) 17–21

[95]   Cocconi G and Morrison P (1959) Searching for interstellar communications. *Nature* **184** 844–6

[96]   Kardashev N S (1979) Optimal wavelength region for communication with extraterrestrial intelligence — $\lambda = 1.5$ mm. *Nature* **278** 28–30

[97]   Mauersberger R *et al* (1996) SETI at the spin-flip line frequency of positronium. *Astronomy & Astrophysics* **306** 141–4

[98]   Kuiper T B H and Morris M (1977) Searching for extraterrestrial civilizations. *Science* **196** 616–21

[99]   The Big Ear site is at http://www.bigear.org/wow.htm

[100] Schwartz R N and Townes C H (1961) Interstellar and interplanetary communication by optical masers. *Nature* **190** 205–8

[101] Betz A (1986) A directed search for extraterrestrial laser signals. *Acta Astronautica* **13** 623–9

[102] For information on optical SETI visit http://www.coseti.org/

[103] The SETI Institute is at http://www.seti-inst.edu/

[104] For information on Project BETA visit http://seti.planetary.org/

[105] For information on Project Argus visit http://seti1.setileague.org/homepg.html

[106] LePage A J (2000) Where they could hide. *Scientific American* **283** (7) 30–1

References

[107] Cohen N and Hohlfeld R (2001) A newer, smarter SETI strategy. *Sky and Telescope* **101** (4) 50–1. See also Hohlfeld R and Cohen C (2000) Optimum SETI search strategy based on properties of a flux-limited catalogue. *SETI beyond Ozma* (Mountain View, CA: SETI Press)

[108] Drake F D and Sagan C (1973) Interstellar radio communication and the frequency selection problem. *Nature* **245** 257–8

[109] Gott J R (1995) Cosmological SETI frequency standards. In Zuckerman B and Hart M H (eds) *Extraterrestrials: Where Are They?* (Cambridge: CUP)

[110] The SETI@home site is at http://www.SetiAtHome.ssl.berkeley.edu/

[111] Lazio T J W, Tarter J and Backus P R (2002) Megachannel extraterrestrial assay candidates: no transmissions from intrinsically steady sources. *Astronomical Journal* **124** (1) to be published

[112] For information on Encounter 2001 visit http://www.encounter2001.com

[113] See http://www.setileague.org/askdr/askdoc.htm

[114] Wigner E (1960) The unreasonable effectiveness of mathematics in the natural sciences. *Communications in Pure and Applied Mathematics* **13** (1).

[115] Budiansky S (1998) *If a Lion Could Talk* (London: Weidenfeld and Nicolson)

[116] Hersh R (1997) *What is Mathematics Really?* (Oxford: OUP); and Dehaene S (1997) *The Number Sense: How the Mind Creates Mathematics* (Oxford: OUP)

[117] Chaitin G J (1997) *The Limits of Mathematics* (Berlin: Springer)

[118] Minsky M (1985) Communication with alien intelligence. In *Extraterrestrials: Science and Alien Intelligence* E Regis (ed) (Cambridge: CUP)

[119] Freudenthal H (1960) *Design of a Language for Cosmic Intercourse* (Amsterdam: North Holland)

[120] Caves C M and Drummond P D (1994) Quantum limits on bosonic communication rates. *Rev. Mod. Phys.* **66** 481–537

[121] Lachman M, Newman M E J and Moore C (1999) The physical limits of communication, or why any sufficiently advanced technology is indistinguishable from noise. http://www.santafe.edu/sfi/publications/wpabstract/199907054

[122] D'Imperio M E (1978) *The Voynich Manuscript — An Elegant Enigma* (Laguna Hills, CA: Aegean Park Press)

[123] Bester A (2000) *The Stars My Destination* (London: Orion). This is a re-issue of the 1956 novel.

[124] Clarke A C (1953) *Childhood's End* (New York: Del Rey)

[125] Tegmark M and Wheeler J A (2001) 100 years of the quantum. *Scientific American* **284** (2) 68–75

[126] Everett H (1957) "Relative state" formulation of quantum mechanics. *Reviews of Modern Physics* **29** 454–62

[127] Turco R P *et al* (1983) Nuclear winter: global consequences of multiple nuclear explosions. *Science* **222** 1283–97

[128] Miller Jr W M (1997) *A Canticle for Liebowitz* (New York: Bantam). This is a re-issue of the original 1959 novel.

[129] Drexler K E (1986) *Engines of Creation: The Coming Era of Nanotechnology* (New York: Doubleday)

[130] Freitas R A Jr (2000) Some limits to global ecophagy by biovorous nanoreplicators, with public policy recommendations. (Available from http://www.foresight.org/NanoRev/Ecophagy.html).

[131] See http://nuclear.ucdavis.edu/rhic.html

[132] See Matthews R (1999) A black hole ate my planet. *New Scientist* 28 August pp 24–7

[133] Jaffe R C *et al* (2000) Review of speculative "disaster scenarios" at RHIC. *Rev. Mod. Phys.* **72** 1125–40

[134] Hut P and Rees M J (1983) How stable is our vacuum? *Nature* **302** 508–9

[135] See http://www.fnal.gov/pub/ferminews/FermiNews98-06-19.pdf

[136] Bird D J (1995) Detection of a cosmic ray with a measured energy well beyond the expected spectral cutoff due to cosmic microwave radiation. *Astrophysical J.* **441** 144–51

[137] Gott III J R (1993) Implications of the Copernican principle for our future prospects. *Nature* **363** 315–319. For a simplified account see: Gott III J R (1997) A grim reckoning. *New Scientist* 15 November pp 36–9

[138] Buch P, Mackay A L and Goodman S N (1994) Future prospects discussed. *Nature* **358** 106–8

[139] Leslie J (1996) *The End of the World* (London: Routledge)

[140] Vinge V (1993) *VISION-21 Symposium* (NASA Lewis Research Center)

[141] Moravec H (1988) *Mind Children* (Cambridge, MA: Harvard University Press)

[142] Searle J R (1984) *Minds, Brains and Programs* (Cambridge, MA: Harvard University Press); Penrose R (1989) *The Emperor's New Mind* (Oxford: OUP)

[143] Knuth D E (1984) *The TEXbook* (Reading, MA: Addison Wesley)

[144] Asimov I (1969) *Nightfall and Other Stories* (New York: Doubleday)

[145] Hart M H (1995) Atmospheric evolution, the Drake equation and DNA: sparse life in an infinite universe. In Zuckerman B and Hart M H (eds) *Extraterrestrials: Where Are They?* (Cambridge: CUP)

[146] Wesson P S (1990) Cosmology, extraterrestrial intelligence, and a resolution of the Fermi–Hart paradox. *Quarterly J. Royal Astronomical Soc.* **31** 161–70

[147] Ward P D and Brownlee D (1999) *Rare Earth* (New York: Copernicus)

[148] Feinberg G and Shapiro R (1980) *Life Beyond Earth* (New York: Morrow)

[149] Asimov I (ed) (1971) *Where Do We Go From Here?* (New York: Doubleday)

[150] Mayr E (1995) A critique of the search for extraterrestrial intelligence. *Bioastronomy News* **7** (3)

[151] Carter B (1974) Large number coincidences and the anthropic principle in cosmology. In *Confrontation of Cosmological Theories with Observation* ed. M S Longair (Dordrecht: Reidel)

[152] Barrow J D and Tipler F J (1986) *The Anthropic Cosmological Principle* (Oxford: OUP)

[153] Livio M (1999) How rare are extraterrestrial civilizations, and when did they emerge? *Astrophysical J.* **511** 429–31

[154] Taylor S R (1998) *Destiny or Chance* (Cambridge: CUP)

[155] For news on extrasolar planets visit http://www.obspm.fr/encycl/encycl.html

[156] Fischer D A *et al.* (2002) A second planet orbiting 47 Ursae Majoris. *Astrophysical J.* **564** 1028–34

[157] McBreen B and Hanlon L (1999) Gamma-ray bursts and the origin of chondrules and planets. *Astronomy & Astrophysics* **351** 759–65

[158] Dole S H (1964) *Habitable Planets for Man* (New York: Blaisdell)

[159] Dole S H and Asimov I (1964) *Planets for Man* (New York: Random House)

[160] Hart H M (1978) The evolution of the atmosphere of the Earth. *Icarus* **33** 23–39; Hart H M (1979) Habitable zones about main sequence stars. *Icarus* **37** 351–7

[161] Kasting J F, Reynolds R T and Whitmire D P (1992) Habitable zones around main sequence stars. *Icarus* **101** 108–28

[162] Gonzalez G, Brownlee D and Ward P D (2001) Refuges for life in a hostile universe. *Scientific American* **285**(4) 60–7

[163] Morbidelli A, Chambers J, Lunine J I, Petit J M, Robert F, Valsecchi G B and Cyr K E (2000) Source regions and timescales for the delivery of water on Earth. *Meteoritics and Planetary Science* **35** 1309–20

[164] Cramer J G (1986) The pump of evolution. *Analog* **106**(1) 124–7

References

[165] Annis J (1999) An astrophysical explanation of the great silence. *J. Brit. Interplanetary Soc.* **52** 19

[166] Asimov I (ed) (1972) *The Hugo Winners, Volumes 1 and 2* (New York: Doubleday)

[167] Harland W B and Rudwick M J S (1964) The great infra-Cambrian glaciation. *Scientific American* **211** (2) 28–36

[168] Hoffman P F and Schrag D P (2000) Snowball Earth. *Scientific American* **282** (1) 68–75

[169] Hoffman P F, Kaufman A J, Halverson G P and Schrag D P (1998) A Neoproterozoic Snowball Earth. *Science* **281** 1342–6

[170] Kirschvink J L (1992) Late proterozoic low-latitude global glaciation: the Snowball Earth. In *The Proterozoic Biosphere* ed. J W Schopf and C Klein (Cambridge: CUP)

[171] Raup D (1990) *Extinction: Bad Genes or Bad Luck?* (New York: Newton)

[172] Gould S J (1986) *Wonderful Life* (New York: Norton)

[173] Alvarez L *et al* (1980) Extra-terrestrial cause for the Cretaceous–Tertiary extinction. *Science* **208** 1094–108

[174] Enever J G (1966) Giant meteor impact. *Analog* **77** (3) 62–84

[175] Alvarez Q (1997) *T-Rex and the Crater of Doom* (Princeton, NJ: Princeton University Press)

[176] Leakey R and Lewin R (1995) *The Sixth Extinction* (New York: Doubleday)

[177] Walker J, Hays P and Kasting J (1981) A negative feedback mechanism for the long-term stabilization of the Earth's surface temperature. *J. Geophys. Res.* **86** 9776–82

[178] Comins N F (1993) *What if the Moon Didn't Exist?* (New York: Harper Collins)

[179] Woese C R, Kandler O and Wheelis M L (1990) Towards a natural system of organisms: proposal for the domains Archaea, Bacteria, and Eucarya. *Proc. Nat. Acad. Sci. USA* **87** 4576–9

[180] Kauffman S (1995) *At Home in the Universe* (London: Viking)

[181] Hartl D L and Jones E W (1998) *Genetics: Principles and Analysis* (New York: Jones and Bartlett)

[182] Watson J D *et al* (1997) *Molecular Biology of the Gene* (4th edn) (Reading, MA: Benjamin/Cummings)

[183] Sorrell H W (1997) Interstellar grains as amino acid factories and the origin of life. *Astrophysics and Space Science* **253** 27–41

[184] Thomas P J, Chyba C F and McKay C P (eds) (1997) *Comets and the Origin and Evolution of Life* (New York: Springer)

[185] Hart M H (1980) N is very small. In *Strategies for the Search for Life in the Universe* (Boston: Reidel) pp 19–25

[186] Clark S (2000) *Life On Other Worlds And How To Find It* (London: Springer/Praxis)

[187] Jakosky B (1998) *The Search for Life On Other Planets* (Cambridge: CUP)

[188] Smith J M and Szathmáry E (1999) *The Origins of Life* (Oxford: OUP)

[189] de Duve C (1995) *Vital Dust* (New York: Basic Books)

[190] Cassen P, Reynolds R T and Peale S J (1979) Is there liquid water on Europa? *Geophysical Research Letters* **6** 731–4

[191] Kovach R L and Chyba C F (2001) Seismic detectability of a subsurface ocean on Europa. *Icarus* **150** 279–87

[192] Fortes A D (2000) Exobiological implications of a possible ammonia–water ocean inside Titan. *Icarus* **146** 444–52

[193] Knoll A H and Carroll S (1999) Early animal evolution: emerging views from comparative biology and geology. *Science* **284** 2129–37

[194] Boesch C and Boesch H (1984) Mental map in wild chimpanzees: an analysis of hammer transports for nut cracking. *Primates* **25** 160–70

[195] Boesch C and Boesch H (1990) Tool use and tool making in wild chimpanzees. *Filia Primatologica* **54** 86–99

[196] Visalberghi E and Trinca L (1989) Tool use in capuchin monkeys: distinguishing between performing and understanding. *Primates* **30** 511–21

[197] Chevalier-Skolnikoff S and Liska J (1993) Tool use by wild and captive elephants. *Animal Behavior* **46** 209–19

[198] Calvin W H (1996) *How Brains Think* (New York: Basic Books)

[199] Gibson K R and Ingold T (eds) (1993) *Tools, Language and Cognition in Human Evolution* (Cambridge: CUP)

[200] Griffin D R *Animal Minds* (Chicago: Chicago University Press)

[201] Ovchinnikov I V *et al* (2000) Molecular analysis of Neanderthal DNA from the northern Caucasus. *Nature* **404** 490–3

[202] Höss M (2000) Ancient DNA: Neanderthal population genetics. *Nature* **404** 453–4

[203] Tattersall I (2000) Once we were not alone. *Scientific American* **282** (1) 56–62

[204] Tattersall I (1998) *Becoming Human* (Oxford: OUP)

[205] Schick K D and Toth N (1993) *Making Silent Stones Speak: Human Evolution and the Dawn of Technology* (New York: Simon and Schuster)

[206] Leakey R (1994) *The Origin of Humankind* (London: Weidenfeld and Nicolson)

[207] Kohn M (1999) *As We Know It* (London: Granta)

[208] Sieveking A (1979) *The Cave Artists* (London: Thames and Hudson)

[209] Quiring R *et al* (1994) Homology of the eyeless gene of Drosophila to the small eye in mice and aniridia in humans. *Science* **265** 785–9

[210] Halder G *et al* (1995) Induction of ectopic eyes by targeted expression of the eyeless gene in Drosophila. *Science* **267** 1788–92

[211] Rogers L J (1997) *Minds of their Own* (Boulder, CA: Westview)

[212] Olson E C (1988) N and the rise of cognitive intelligence on Earth. *Quarterly J. Royal Astronomical Soc.* **29** 503–9

[213] Pinker S (1994) *The Language Instinct* (London: Allen Lane)

[214] Lai C S L *et al* (2001) A forkhead-domain gene is mutated in a severe speech and language disorder. *Nature* **413** 519–23

[215] Asimov I (1984) *Asimov's New Guide to Science* (New York: Basic Books)

[216] Lines M E (1986) *A Number For Your Thoughts* (Bristol: Adam Hilger)

[217] Gould S J (1996) *Life's Grandeur* (London: Cape)

[218] Monod J (1971) *Chance and Necessity* (London: Collins)

Index

Columbus Optical SETI Observatory, 101
Columbus, C., 54
comets, 162, 268, 269
Compton, A. H., 242
Conant, J. B., 242
Contact, 69, 101, 258
Cook, J., 54
Copernican principle, 133
Copernicus, N., 3, 61
cosmic rays, 64, 92
Cramer, J. G., 164, 165, 169
Crawford, I. A., 74, 254
Cretaceous extinction, 176, 177
Crick, F. H. C., 45, 46, 81, 250, 272
Cro-Magnon man, 217
crop circles, 32
Cros, C., 260
Cryptozoic era, 176
cyanobacteria, 204
cytosine, 195, 272
cytoskeleton, 208

d'Arrest, H. L., 39, 249
Darby, A., 232
Darwin, 110
Darwin, C., 57
Davis Jr., R., 92, 257
Davis, D. R., 271
Deardorff, J. W., 47, 48, 250
Deimos, 189, 249
Deinococcus radiodurans (Conan the Bacterium), 124
delta *t* argument, 130–134
Devonian extinction, 176
DiPietro, V., 249
directed panspermia, 45, 46, 63, 81, 83
Dixon, P., 129
DNA, 43, 124, 192–202, 208
Drake equation, 19, 20, 107, 134, 141, 150, 152, 218, 235
Drake, F. D., 20, 91, 98, 99, 102–104, 106, 107, 110, 111, 115, 161, 245, 261
Drexler, K. E., 263
Dr. Strangelove, 261
drunkard's walk, 240
Dumas, S., 260
Dunn, A., 17
Durand de Saint-Pourçain, G., 247
Dutil, Y., 260
Dyson sphere, 90, 106, 256
Dyson, F. J., 45, 67, 84, 253

Earth, 35–37, 161, 184
  Early Archean era, 204
  Hadean era, 204
  magnetic field, 181
  obliquity, 188
  plate tectonics, 181–183
  temperature, 183
  water oceans, 163
Earth–Moon system, 36, 37, 185–187
Easter Island, 34
Einstein, A., 9, 68, 244, 257
Ekbom, L., 243
enzymes, 191, 197
Epsilon Eridani, 63, 64, 98, 99, 102, 161
Eratosthenes of Cyrene, 235, 274
Escher, M. C., 1
ET, 258
eubacteria, *see* bacteria
Eubulides of Miletus, 13, 243
eukaryotes, 179, 190, 191, 193, 199, 207, 208
Europa, 205
Everett III, H., 262
Evpatoria telescope, 260
eye (evolution of), 222
eyeless gene, 222, 223

Farmer, P. J., 261
fata bromosa (mirage), 30
fata morgana (mirage), 30
Feinberg, G., 247
Fermi, E. (biographical details), 7, 8, 10, 17, 20–22, 28, 29, 242, 245
Fermi, L., 9
Fermilab, 129, 130
fermions, 8
flying saucers, *see also* UFOs, 17, 29, 30, 33, 47
Fogg, M. J., 49, 50, 74, 82, 251, 254
Forbidden Planet, 1, 241
47 Ursae Majoris, 154, 155
Forward, R. L., 66, 253
Foster, J., 253
Freitas Jr., R. A., 1, 2, 127, 241
Freudenthal, H., 119, 262
FTL travel, 67

Gabor, D., 246
Galactic Club, 49, 51
Galilei, Galileo, 231
gamma-ray bursters, 101, 106, 130, 157, 167, 170–172

Index